Lecture Notes in Geosystems Mathematics and Computing

Lecture Notes in Geosystems Mathematics and Computing series showcases topics of current interest that lie at the interface of the geosciences, mathematics, and observational and computational sciences. Titles may present new results or novel perspectives on existing areas of research, or serve as extended surveys of certain topics. Traditional lecture notes covering core concepts for graduate and doctoral students are also included, as are titles that are based on expansions of outstanding PhD theses. Topically, titles address either mathematical and computational concepts and techniques – such as inverse problems, applied harmonic analysis, numerical simulation, and data analysis – or focus on practical applications in such areas as gravitation, climate modeling, seismology, geomechanics/dynamics, and satellite technology.

All manuscripts are peer-reviewed to meet the highest standards of scientific literature. Interested authors may submit proposals by email to the series editors or to the relevant Birkhäuser editor listed under "Contacts."

More information about this series at http://www.springer.com/series/15481

Christian Blick • Willi Freeden • M. Zuhair Nashed
Helga Nutz • Michael Schreiner

Inverse Magnetometry

Mollifier Magnetization Distribution
from Geomagnetic Field Data

Christian Blick
Business und Management mbH
CBM - Gesellschaft für Consulting
Bexbach, Saarland, Germany

Willi Freeden
Mathematics Department
University of Kaiserslautern
Kaiserslautern, Rheinland-Pfalz, Germany

M. Zuhair Nashed
Mathematics Department
University of Central Florida
Orlando, FL, USA

Helga Nutz
Business und Management mbH
CBM - Gesellschaft für Consulting
Bexbach, Saarland, Germany

Michael Schreiner
OST
Institute for Computational Engineering
Buchs, Switzerland

ISSN 2730-5996   ISSN 2512-3211 (electronic)
Lecture Notes in Geosystems Mathematics and Computing
ISBN 978-3-030-79507-8   ISBN 978-3-030-79508-5 (eBook)
https://doi.org/10.1007/978-3-030-79508-5

Mathematics Subject Classification: 35R25, 35R30, 65N21, 86A22

This book is published under the imprint Birkhäuser, www.birkhauser-science.com by the registered company Springer Nature Switzerland AG
The registered company address is: Gewerbestrasse 11, 6330 Cham, Switzerland

Preface

The different methodological facets of potential theory have changed considerably in the last decades. New applications in geoengineering have led to changed aspects in potential theory, to well-posed boundary value problems, for example, in physical geodesy and geophysics, and also to ill-posed inverse problems in exploration. A particularly significant problem in exploration besides seismics and gravimetry is *inverse magnetometry* in its characteristics of a dipole-based framework. It is well known that inverse magnetometry may be described by a singular Fredholm integral equation of the first kind relating the vectorial magnetization distribution to data of the geomagnetic field. Several "solution proposals" exist based on standard regularization procedures. However, in the opinion of the authors of this work, inverse magnetometry requires a special "torture" in the context of the regularization methodology of the theory of inverse problems. The reason is that inverse magnetometry is "too ill-posed" to be solved with a standard regularization method. This opinion has considerable consequences for the approach conveyed in this work: It leads to the so-called "mollifier technique in a dipole oriented context," that is the central research topic of this book. In doing so, larger parts of the book considerably parallel the monopole mollifier theory known from inverse gravimetry, but other parts inextricably demand their own dipole-based settings and concepts.

The specific objective for the book is to write a geomathematical contribution that would be a companion to the work "W. Freeden [45], Decorrelative Mollifier Gravimetry. Geosystems Mathematics, Birkhäuser, 2021." The final outcome of our work presented under the title "Inverse Magnetometry: Mollifier Magnetization Distribution from Geomagnetic Field Data" is not a textbook; it is a research monograph. Notably, the research intention of the authors from the beginning was to fill up a scientific niche in mathematics, where it was envisioned that the book should provide a comprehensive coverage of application of the mollifier methodology to magnetometry.

The book is dedicated to near-surface as well as deep geology with geomagnetic potential input data primarily of terrestrial origin, but the geomathematical decorrelation methods are to be designed in such a way that depth information (e.g., in boreholes) may be canonically entered. As for the monograph [45], we are allowed to state that, starting from an appropriately dense and accurate data set, mollifier wavelet methods are able to specify with increasing scales the interfaces (transitional surfaces) of geological strata and the locations of fault structures, however, without intrinsically revealing the specific rock type of the respective geological formation. All in all, the book pursues a double goal: On the one side, it represents a scientific set of rules for today's geoengineering, interested in the application of innovative mathematical modeling and simulation techniques to promising data sets and structures. On the other side, the book serves as a collection of current material in Applied Mathematics to offer alternative methodologies in the theory of inverse problems.

May 2021   C.B., W.F., M.Z.N., H.N., M.S.

Goal of the Book

The *main goal of this work* is to demonstrate that the key technology of geo-mathematics in the particular manifestation of mollifiers is, indeed, capable of reducing inverse magnetometry to simple decorrelation models that are accessible for exploration purposes. More than that, our contribution familiarizes a broad audience with the diverse innovative questions and problems of today's mollifier magnetometry by closely parallelizing the ideas and concepts developed in the book [45] about decorrelative mollifier gravimetry. In the case of inverse magnetometry, the mollifier methods, in fact, lead to the result that we are able to specify the interfaces (transitional surfaces) of geological strata with increasing scales by means of mollifier wavelet techniques, without intrinsically revealing the specific rock type of the respective geological formation. For the geological approach of the areas between the interfaces either extrapolating experience of a geologist based on *a priori* knowledge (e.g., information from previous drilling activities in neighboring areas and knowledge from previous mining use) or *a posteriori* validation, for example, by drilling, is required.

To be more concrete, this book presents an insight into the current state of research of reducing magnetometry to mathematically accessible and thus calculable decorrelated models. In this way, the various geomathematical problems of magnetometry are made available to a broad scientific audience and the exploration industry. New impulses are given and innovative ways of modeling geologic layers by mollifier techniques are shown.

All in all, the purpose of the book is to provide pioneering and ground-breaking innovative mathematical knowledge as a transfer methodology from the "reality space" of geomagnetic measurements and observations into the "virtual space" of mathematical-numerical modeling structures and mollifier solutions with novel geological application areas. The essential significance of the book is the bridging function of geomathematics as key technology, in many ways: The book leads in a cycle from the potential measurements by geoengineers and geophysicists, the subsequent theory and model formation, computer-based implementation, and numerical calculation by geomathematicians to interpretation by geologists. It therefore spans the spectrum from geoengineering via geophysics to geomathematics and geology, and back if necessary.

Structure of the Book

The book presents the context arising in the mathematical handling of decorrelative magnetic exploration:

- The introductory remarks in Chap. 1 are concerned with geomathematically oriented aspects about the history and the methodology of Earth's magnetic field research.
- Chapter 2 characterizes the ingredients of the Earth's magnetic field, but only to the extent needed for the book. An insight into the constituents of crustal geomagnetic field research is given.
- Chapter 3 recapitulates results of potential theory as far as they are significant for the understanding of dipole potential based magnetometry. The various questions and problems of dipole-oriented magnetometry are made scientifically available.
- Chapter 4 discusses inverse magnetometry as an ill-posed problem in a dipole reflected nomenclature. It is mentioned that all criteria of Hadamard's classification (existence, uniqueness, and stability) are violated for terrestrial data. Consequently, inverse magnetometry is considered to be "too ill-posed" in order to use standard regularization techniques other than mollifier regularization.
- Chapter 5 explains the origin of mollifier approximation; the mollifier philosophy is explained in more detail; and new impulses are given to the theory of inverse problems by mollifier techniques. The specific role of mollifier approximation is pointed out within the multi-scale context using mollifier dipole potentials corresponding to Haar-type Dirac sequences. Haar-wavelet based decorrelation of both magnetization distribution and geomagnetic potential data are investigated. The transfer methodology is provided from geomagnetic measurements to space-limited mollifier modeling of geological structures as well as mollifier reconstruction and decomposition. Altogether, innovative ways of modeling geologic layers by mollifier techniques are developed by reducing magnetometry in a multi-scale concept to mathematically accessible and thus calculable decorrelated models.
- Chapter 6 studies graphical demonstrations of locally supported Haar mollifier based decorrelation for special test examples, namely the Marmousi-model and an area of the Bavarian Molasse Basin.
- The concluding remarks in Chap. 7 describe some fundamentals of the spectrum of mollifier techniques applicable to future subsurface geology.

Acknowledgments

The CBM-authors thank the "Federal Ministry for Economic Affairs and Energy, Berlin" and the "Projektträger Jülich (PtJ)" (scientific managers Dr. V. Monser, Dr. S. Schreiber) for funding the project "SYSEXPL" (PtJ funding reference number: 03EE40002A, PI Prof. Dr. W. Freeden, University of Kaiserslautern, Germany, corporate manager Prof. Dr. M. Bauer, CBM – Gesellschaft für Consulting, Business und Management mbH, Bexbach, Germany). The book is intended as a collection of mathematical research tools that should be internalized by researchers in order to successfully approach a mollifier research project of geo-exploratory practice. For the PtJ-project SYSEXPL it is to be considered as a mathematical background, but without representing any parts of the numerical implementation and geo-practically relevant results to be done there for the German country Saarland. The outcome of specific material of the PtJ-research project SYSEXPL is reserved for future publications. All in all, the contribution about decorrelative mollifier magnetometry as presented here is a canonical mathematical continuation of the work already described in W. Freeden, Decorrelative Mollifier Gravimetry: Basics, Concepts, Examples and Perspectives, Geosystems Mathematics, Birkhäuser, Basel, 2021 (see [45]), from which some parts about magnetometry are literally taken over. Moreover, the monograph [45] published 2021 by W. Freeden is in its application part an English translation of the German book by W. Freeden, M. Bauer [46] published 2020 in the "Springer Spektrum" series. The research of Michael Schreiner has been supported by the Swiss Federal Office of Energy, under grant agreement SI/501874-01.

The authors would like to thank Springer for the opportunity to publish this geomathematical research book.

Contents

Chapter 1
Introductory Remarks

An essential objective of mathematics is to create settings and concepts to better understand our world. Mathematics is present in everyday life. Even more, almost all sciences undergo a process of "mathematization" due to increasing technological progress. Mathematics as a cross-sectional science permeates almost all areas of our life and society. As a result, mathematics is in close interaction with the Earth, nature, environment, technical and economic sciences, medicine and parts of the humanities (*"Mathematization of Sciences"*).

Today, the use of the computer enables us to treat complicated models to real data sets. Modeling, calculation and visualization lead to reliable simulations of processes and products. Mathematics is the "raw material" of the models and the essence of every computer simulation; it forms the mediator (i.e., the key technology) to convert the images of the real world into models of the virtual world, and vice versa (cf. Fig. 1.1).

1.1 Cycle of Measurement and Mathematical Modeling

Geomathematics (cf. [43, 44, 70]) is devoted to the qualitative and quantitative properties of the currently existing or possible structures of our Earth system. Geomathematics is the pate of the concept of scientificity in Earth system research. The Earth system consists of a number of systems, which themselves represent a system. The complexity of the overall Earth system is determined by interacting physical, biological, and chemical processes that transform and transport energy, material, and information. It is characterized by interdependencies of natural processes with social and economic processes that lead to mutual interaction.

The Earth, in fact, is a prime example of a *complex system* with interacting natural and socio-economic subsystems. In the consequence, a simple cause-effect thinking is completely unsuitable for an appropriate understanding. What is usually

© The Author(s), under exclusive license to Springer Nature Switzerland AG 2021
C. Blick et al., *Inverse Magnetometry*, Lecture Notes in Geosystems Mathematics and Computing, https://doi.org/10.1007/978-3-030-79508-5_1

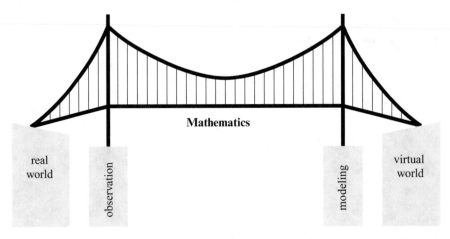

Fig. 1.1 Mathematics as a key technology bridging the real and virtual world (illustration taken in modified form from [66])

required is thinking in awareness of effect-cause structures with certainly sometimes undesirable effects in observation and modeling.

All the aspects described above indispensably demand a mathematics, which must be more than a collection of theories and numerical procedures. Rather, geomathematics is nothing more than an organization of the complexity of the Earth system. This includes descriptive thinking to clarify abstract complex issues, proper simplification of complicated interactions, an appropriate mathematical system of concepts for description, and accuracy in thinking and formulating. Geomathematics thus becomes the key science of the complex Earth system.

Wherever there are data and observations, e.g., with the various scalar, vectorial, and tensor clusters of satellite data or terrestrial gravitational field and/or magnetic field data, it becomes mathematical. Statistics serves for denoising, constructive approximation is concerned with compression, special systems of functions give georelevant graphical and numerical representations, all this with mathematical algorithms.

From our comments, it becomes clear that geomathematics as the mathematics dedicated to the Earth sciences becomes increasingly important also in the field of exploration. It has the special task of building a bridge between mathematical theory and geotechnical application. The special attraction of geomathematics is therefore based on the lively exchange of ideas between on the one hand the more in modeling, theoretical foundation and approximate and numerical problem solving interested group of applied mathematicians and on the other hand the more with measurement technology, methodology of data analysis, implementation of routines and software application familiar group of geoengineers.

To the present day, computers and measurement technology have led to an explosive spread of mathematics, in general. As a result, mathematics is in close interaction with the Earth sciences (cf. Fig. 1.2).

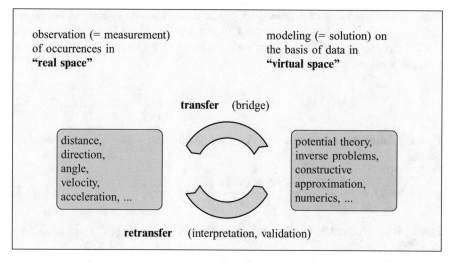

Fig. 1.2 Circuit "real space" (observation) and "virtual space" (modeling) in modified form following W. Freeden (2018), Introduction to the "Handbook of Mathematical Geodesy" [69]

All the aforementioned generally valid aspects explain why Geomathematics is a key technology for *potential methods in exploration*, in particular for magnetometric engagement.

1.2 Specific Problems in Inverse Potential Field Recovery

The fact that the geomagnetic field is usually measured from outside the object Earth rather than from inside means that the description of the object or parts of it is inferred from the geomagnetic field by using the measurements to specify a limited number of parameters of a model that might describe the object. This process is known as the *"inverse magnetometric problem"* and involves obtaining a description of the magnetization distribution from measurements of the magnetic field.

The easiest example of the geomagnetic inverse problem (see, e.g., [94]) is to determine the average remanent magnetization of a large spherical object from a magnetic field measurement. If the sphere was known to be homogeneously magnetized, the magnetic field outside of the sphere would be identical to that of a point magnetic dipole located at the center of the sphere. The magnetic field b at the point x produced by a point magnetic dipole m at the point y is given by

$$b(x) = \frac{\mu_0}{4\pi} \left(\frac{3 \, m(y) \cdot (x - y)}{|x - y|^5}(x - y) - \frac{m(y)}{|x - y|^3} \right), \qquad (1.1)$$

where μ_0 is the vacuum permeability. Since this equation is linear in the dipole moment $m(y)$, if the location y of the dipole is known, it can be inverted to obtain the three components of $m(y)$ from the knowledge of the three components of b at a single point x

$$m(y) \cdot (x - y) = \frac{2\pi}{\mu_0} |x - y|^3 \, b(x) \cdot (x - y). \tag{1.2}$$

By substituting this expression into (1.1), we obtain after some elementary manipulation

$$m(y) = \frac{4\pi}{\mu_0} |x - y|^3 \left(\frac{3}{2} \frac{b(x) \cdot (x - y)}{|x - y|^2} (x - y) - b(x) \right). \tag{1.3}$$

Hence, it is sufficient to measure the magnetic field b at a single known location x outside of the sphere.

However, if our sphere has only a small number of regions that are magnetized but at unknown locations within the sphere, we would need multiple dipoles to describe the field. Since the location of these dipoles is unknown, the non-linearity of (1.1) in x and y makes the inverse process much harder; in general, there is no closed-form analytical solution for determining m from measurements of b at multiple locations.

In this context, there is another serious implication of (1.1): the fall-off of the field with distance serves as a harsh spatial low-pass filter, so that the further a magnetic object is from the measurement location, the greater is the spatial blurring of the contribution of adjacent source regions. The loss of information with distance is so rapid that it often cannot be balanced by realistic reductions of sensor noise. More importantly, if the object we are studying contains a spherical shell of uniform radial magnetization, the integration of (1.1) over that shell would produce a zero magnetic field outside of the shell. Hence, no magnetic measurements and inverse process would be able to detect the presence of such a closed shell which is located somewhere inside the object.

Similar problems occur in the interpretation of magnetic fields from current sources in conducting objects, whether they are a heart, a brain, or a corroding aircraft wing: Whenever a measured field obeys Laplace's equation, there exists the possibility of source distributions with symmetries such that they produce no externally detectable fields. The ability to add or subtract such silent sources without altering the measured field corresponds to the lack of a unique solution to an inverse problem.

More generally, inverse potential field problems (IPFPs) such as inverse magnetometry arise in many branches of science and mathematics, including geostatistics, medical imaging (such as tomography), non-destructive testing, geosystems mathematics (especially involving, e.g., gravimetric, geomagnetic, and seismic exploration and satellite observational technology), and many other fields. IPFPs are confronted with the invariable contamination of potential data due to measurements.

The rationale in most methods for resolution (i.e., the "approximate solvability") of an inverse potential field problem IPFP is to construct a "solution" that is acceptable physically as a meaningful approximation and is sufficiently stable from the computational standpoint.

The main dilemma of "solving" inverse problems such as IPFPs is that they are ill-posed, i.e., existence, uniqueness as well as continuous dependence criteria are violated. The characteristic of such problems is that the closer the mathematical model describes the ill-posed problem (IPP), the worse is the "condition" of the associated computational problem (i.e., the more sensitive to errors).

Therefore, the indispensable problem is to bring additional information about the desired solution, compromises, or new outlooks as aids to the resolution of ill-posed problems (IPPs). It is conventional to use the phrase "regularization of an ill-posed problem" to refer to various approaches to circumvent the lack of continuous dependence as well as to bring about existence and uniqueness if necessary. Roughly speaking, this entails a treatment of an IPP via an analysis of an associated family (usually a sequence or a net) of well-posed problems (WPPs), yielding meaningful answers to the IPP.

It should be remarked that IPPs are a very active field of research in applied mathematics, with a fastly growing bibliography (see, e.g., [56, 117, 119–122]). Throughout this contribution about inverse magnetometry, we shall refer to various papers and monographs including further details on several aspects. Gravimetric and magnetometric IPFPs, however, are not usually associated with standard regularization techniques. They are "too strongly ill-posed" to apply one of the traditional regularization methods, so that a mollifier approach should be attempted. As a matter of fact, the original idea of the mollifier method in the general theory of inverse problems is already found, e.g., in the work published by Engl et al. [36], Louis and Maass [95]. The heuristic motivation for the "mollification" in these early contributions is, in particular, the annoyance that high-frequency components in the solution must be specifically damped by smooth structures in order to obtain solution statements at all.

The decorrelation concept of wavelet mollifiers as presented in this contribution is a continuation of earlier approaches, which already include the intuitive idea of regularization through a "mollification" (mollifier regularization) over a surface (cf. [64, 65]) or a volume (cf. [47]). In fact, following [47, 64], the (singular) fundamental kernel of the "2D-Beltrami (surface) operator" and the negative "3D-Laplace (volume) operator," respectively, are replaced by suitable mollifier sequences of regular kernels in close physical proximity (see also the studies in [49, 58–60]). Decorrelative exploration was numerically tested in the PhD theses [19, 40, 113, 147] of the Geomathematics Group, University of Kaiserslautern and recently by the CBM-research group, Bexbach, Saarland, Germany.

Finally, it should be remarked that inverse magnetometry as a particularly important example of the theory of inverse problems represents a central research object in geophysics and geoexploration. There is a large number of inversion techniques, which we only mention for introduction (e.g., [18, 24, 25, 31, 38, 39, 49, 90–92, 98, 109, 137, 145, 146, 148] and the literature listed therein).

Chapter 2
Basics of Magnetic Field Theory and Magnetization

Like the gravitational field, the magnetic field is a vector field. Gravity and magnetic exploration techniques are both demanding in that they exploit naturally existing fields of the Earth. The measured quantities are integrated effects of the subsurface and need to be decorrelated in order to offer interpretability.

2.1 Magnetometric Methods: Historical Stages

Gravity and magnetic field inversion leading to density and magnetization distribution, respectively, may be described both by inverse problems of potential theory. Both problems may be formulated by a Fredholm integral equation of the first kind. Both problems lack well-posedness, from which the non-uniqueness is a dramatic feature. So, there is a long history to apply gravity and magnetic methods and tools in parallel.

However, there are also considerable differences: The mass density distribution in Newtonian gravitational theory is a scalar function, while the material magnetization distribution is of vectorial nature. The gravitational field points to the center of mass of the Earth, while the magnetic field may be understood as similar to a bar magnet at the Earth's center (cf. Fig. 2.3).

In order to understand magnetic fields for specific purposes of geoexploration, an explanation of the history of geoscientific laws that govern these fields is useful. Some milestones on the way to current potential methods of magnetometry are in the following list, which, however, does not reflect the huge modern developments during the last decades in theoretical and practical perspective (cf. [45, 46]):

ca. 1600 W. Gilbert (1544–1608): Founder of the doctrine of Earth magnetism,

ca. 1750 B. Franklin (1706–1790): Magnetized sewing needles due to Joule heating in the geomagnetic field,

© The Author(s), under exclusive license to Springer Nature Switzerland AG 2021
C. Blick et al., *Inverse Magnetometry*, Lecture Notes in Geosystems Mathematics and Computing, https://doi.org/10.1007/978-3-030-79508-5_2

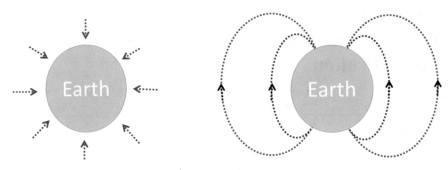

Fig. 2.3 The gravitational field lines are monopolar, i.e., they point to the center of mass of the Earth (left), while the magnetic field lines are dipolar (right, see Fig. 3.8 for a more precise illustration)

ca. 1780	A.M. Legendre (1752–1833), P.S. Laplace (1749–1827): Development of spherical harmonics (multipoles),
ca. 1795	C.A. de Coulomb (1736–1806): Finding that magnetic forces obey the inverse square law,
ca. 1820	H.C. Oerstedt (1777–1851): Influence of electric currents on a magnetic needle (beginning of electromagnetism),
ca. 1820	J.B. Biot (1774–1862), F. Savart (1791–1841): Generation of a magnetic field by electric currents ("Biot–Savart Law"),
ca. 1835	C.F. Gauss (1777–1855): Beginnings of potential theory, magnetic field developments by means of spherical harmonic expansions, electric currents in upper atmosphere (as hypothesis),
ca. 1865	J.C. Maxwell (1831–1879): Dynamical theory of the electromagnetic field,
ca. 1878	B. Stewart (1851–1935): Basic idea of the dynamo theory in magnetics,
ca. 1907	A. Schuster (1851–1934): First quantitative dynamo theory,
1929	O.D. Kellogg (1879–1932): Standard textbook on classical potential theory (his doctoral advisor was D. Hilbert in Göttingen),
since 1980	Wavelet methods in Geophysics and Geomathematics.

This incomplete list, essentially related to the history of geoscientific exploration up to the last century, already explains that proper geomathematical understanding is a key factor in exploratory research.

There are no magnetic monopoles (in contrast to gravitation) and, hence, dipoles (and higher order poles such as quadrupoles aso.) are the principal constituents of magnetic fields. There is also another fundamental difference: magnetic poles can be repulsive or attractive, they are not always attractive as in the gravity case. As a consequence, the geophysically close construction of gravimetrically and magnetometrically reflected wavelets has to reflect the monopole and dipole character of the potentials to be investigated. Therefore, the mathematical treatment of decorrelation methods in gravimetry and magnetometry is different. Moreover, the theoretical and methodological technicalities in decorrelative gravimetry are much less than in decorrelative magnetometry. In fact, the numerical aspects of decorrelative magnetometry are still in the developmental stage, and the interpretability of the results obtained so far is less meaningful than in gravimetry. The present contribution addresses these shortcomings with the aim of improving the situation in the near future.

2.2 Essential Constituents of the Earth's Magnetic Field

As already pointed out, magnetometry, being another potential method, has a lot of similarities with gravimetry and, hence, they are often measured and interpreted together. A decorrelation methodology similar to that of gravimetry is also feasible. However, there are also significant differences. As already mentioned, there are no magnetic monopoles (in contrast to gravity) and, hence, dipoles (and sometimes higher order poles such as quadrupoles aso.) are principal constituents. The magnetic field of the Earth is also less stable than the gravity field. It is changing quickly. In contrast to gravity maps, the magnetic maps are dominated more essentially by local anomalies. The differences in magnetization of different rock types are often quite large (much larger differences than in the case of densities).

The magnetizing field strength is determined, following the Biot–Savart law (see e.g., [18, 114]), as being the field strength at the center of a loop of wire through which a particular current is flowing. The magnetic flux lines between two poles per unit area characterize the flux density and are measured in tesla, one tesla named in honor of N. Tesla (1856–1943) is equivalent to 10^4 G (Gauss), used in the CGS system, $1 \, \text{kg} \, \text{s}^{-2} \text{A}^{-1}$ used in SI-base units. So, tesla is a derived unit of the magnetic induction (also, magnetic flux density) in the *International System of Units*, which can also be expressed as

$$T = \frac{Vs}{m^2} = \frac{N}{Am} = \frac{J}{Am^2} = \frac{HA}{m^2} = \frac{Wb}{m^2} = \frac{kg}{Cs} = \frac{Ns}{Cm} = \frac{kg}{As^2}, \tag{2.4}$$

where A = ampere, C = coulomb, kg = kilogram, m = meter, N = newton, s = second, H = henry, V = volt, J = joule, and Wb = weber (for more details the reader is referred to textbooks about electric and magnetic fields).

The unit of tesla [T], however, is too large to be practical in exploration, so a subunit called a nanotesla ($1 \, \text{nT} = 10^{-9}$ T) is used instead (where 1 nT is equivalent to 10^{-5} G).

Our very brief insight into the Earth's magnetic field presented in the following is taken almost literally from the monographs [45, 49]. Recent extensive treatments are gathered in [75, 77, 88]. A comprehensive description with focus on the lithospheric field and satellite data processing is given in [89]. Some shorter overviews are, e.g., in [83, 84, 127].

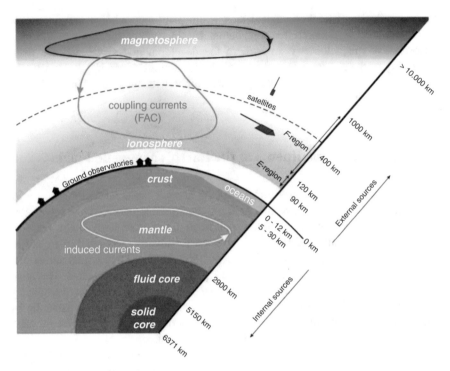

Fig. 2.4 Schematic description of the contributions to the Earth's magnetic field (Courtesy of [127])

The *(primary) geomagnetic field* denotes the magnetic field generated by all sources inside and outside the solid Earth up to the magnetopause. The magnetopause forms the transition layer between the geomagnetic field and the interplanetary magnetic field (IMF) originated from solar processes. The Earth's magnetic field can be divided into the following major source regions (see Fig. 2.4):

- *Core.* Convection in the Earth's liquid core drives dynamo processes that generate by far the largest part of the geomagnetic field (with a field strength varying between 30,000 nT and 60,000 nT at the Earth's surface). Thus, the core field is sometimes also called the *main field.* It has a dominating dipole component and is of rather large scale, concerning its spatial and temporal variation.

- *Crust/Lithosphere.* The Earth's crust and mantle produce a further contribution to the magnetic field, called the *secondary (or anomalous) field.* Magnetization of minerals can only take place at temperatures below the specific Curie temperature, so that the crustal field has its sources in depths no more than a few tens of kilometers below the Earth's surface. The magnetization can be remanent (i.e., it has taken place in the past) or induced by an ambient magnetic field (in first place, the core field). The magnetic signature of such magnetized minerals reveals a strong spatial variation and can reach from a few nanoteslas to up to over 1000 nT (locally). Together with the core field, the crustal field forms the so-called *internal field*.

- *Ionosphere*. The ionosphere denotes approximately the region between 50 km and 1000 km above the Earth's surface, where solar heating leads to higher conductivity on the dayside than on the nightside and drives different electric current systems. An example is the eastward directed equatorial electrojet (EEJ), which is due to an enhanced conductivity along the dip equator, while an enhanced conductivity in the polar regions drives the auroral electrojet (AEJ). These current systems produce additional magnetic fields contributing to the geomagnetic field. A further permanently present magnetic field is the so-called solar-quiet (Sq) variation of about 20−50 nT (which can be significantly stronger during "magnetically disturbed" times). The ionospheric field in general shows strong spatial and temporal variations.

- *Magnetosphere*. The magnetosphere extends beyond the ionosphere up to the magnetopause. Due to deformation by the solar wind, it has an extent of few tens of Earth's radii on the dayside, but reaches up to several hundred Earth's radii on the nightside. Major current systems are the magnetopause current, the magnetotail current, and the equatorial ring current. Their magnetic effect is spatially of rather large scale but can vary significantly in time. A coupling with the ionosphere is established by field-aligned currents (FAC) that flow along the core field lines and can be detected at satellite altitude as non-potential magnetic fields. Together with the ionospheric contribution, the magnetospheric magnetic field forms the *external field* (see Fig. 2.5).

The description above gives a first impression of the complexity of the Earth's magnetic field (see also [76, 97, 141]). The crustal field has hardly any noticeable time dependence, and the core field only shows long term variations in time. The secular variation, i.e., the time change of the core field, is of the order of 50 nT per year. Occasional changes at shorter timescales with an increased rate of secular variation are the so-called geomagnetic jerks. At larger timescales of several thousand or million years, a complete reversal of the direction of the main field becomes possible. Paleomagnetic records show that this has happened many times during the Earth's life span. Changes of the external field take place on significantly smaller timescales. There are the daily variations due to the different solar influence on the day- and night-side, variations due to changes in the interplanetary magnetic field, as well as disturbances by magnetic storms or substorms due to an increased solar activity. The initial phase of a magnetic storm usually lasts only a few minutes, while the main phase can extend up to several hours, and the subsequent recovery phase up to a few days. Amplitudes of such storms are generally about a few tens of nanoteslas but can reach more than 1000 nT. In the year 1989, a particularly strong magnetic storm caused the shutdown of parts of the Canadian electrical power grid.

Concerning the spatial extent, core field and magnetospheric field change at significantly larger scales (of several thousands of kilometers) than the ionospheric and the crustal field. Spatial scales for the crustal field variability are often below 500 km. A typical and well-known example is the Bangui anomaly in central Africa.

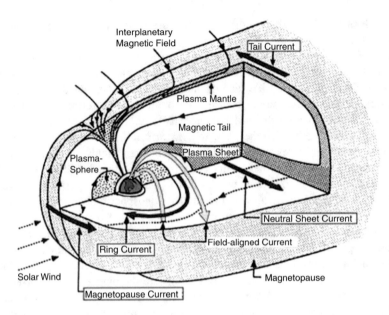

Fig. 2.5 Illustration of current systems in the magnetosphere and their coupling with the ionosphere (cf. [49])

Actually, the crustal field can reveal variations on severely smaller spatial scales. Concentrations of magnetic ores can produce magnetic fields of up to several thousand nanoteslas in an area of tens of meters. The detection of such local variations, however, requires terrestrial or aeromagnetic measurements. With all these contributions, it is a tremendous effort to achieve a good description of the geomagnetic field from the available data (spaceborne (satellite), airborne (aeromagnetic), and terrestrial measurements). The different spatial and temporal scales make it difficult to find appropriate modeling approaches–yet these differences allow the separation of the different contributions at least too long to a certain degree.

A study by Mauersberger [105] and Lowes [96] of the power (spherical harmonics) spectrum of the internal contributions (cf. Fig. 2.6) shows a significant change in the (orthogonal) spherical harmonic expansion around the spherical harmonic degree 15, which is interpreted as the transition from the large-scale core field dominated part to the small-scale crustal field dominated part of the magnetic field (cf. [126]). This is still today the (empirical) criterion to truncate models of the main field globally.

Also the choice of the coordinate system (see, e.g., [127]) can improve the understanding of the different contributions. While the internal field is modeled in an Earth-fixed coordinate system using geographic longitude and colatitude, the external field is usually modeled in a sun-fixed coordinate system, e.g., using local

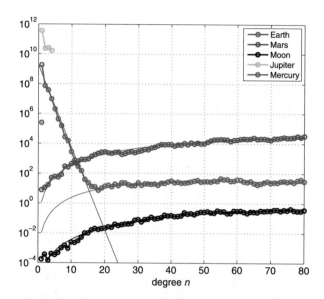

Fig. 2.6 Mauersberger–Lowes power spectra in terms of spherical harmonics of the internal field of the Earth (Courtesy of [127]). In addition, the theoretical crustal spectra (thin curves) for the Earth, Mars, and the Moon are shown. Note that the lack of any significant core field in the case of Mars and the Moon displays pure crustal types of spectra. Furthermore, magnetic fields of planets may be very weak (e.g., Venus) or much stronger than the Earth's field (e.g., Jupiter). The magnetic dipole moment of Jupiter is about 10^5−times higher than that of the Earth

time and dipole latitude (thus, paying tribute to the dominating solar influence for the external field).

Relevant PhD theses of the Geomathematics Group, University of Kaiserslautern, in geomagnetic research are [14, 74, 99, 106]. Further publications following the concepts of the Geomathematics Group Kaiserslautern are [15, 16, 48, 49, 52–54, 100, 108].

2.3   Maxwell Equations and Law of Biot–Savart

Next we focus on mathematical tools in geomagnetic modeling that are of importance in potential methods in exploration.

The fundamental relations of any electrodynamic process are described by the full set of Maxwell equations:

$$\nabla \wedge e = -\frac{\partial}{\partial t}b, \tag{2.5}$$

$$\nabla \cdot d = \rho, \tag{2.6}$$

$$\nabla \wedge h = j + \frac{\partial}{\partial t} d, \tag{2.7}$$

$$\nabla \cdot b = 0, \tag{2.8}$$

with b denoting the magnetic induction, e the electric field, j the density of free currents, and ρ the density of free charges.

Further relationships between the magnetic (displacement) field h and the electric displacement field d are of the form

$$d = \varepsilon_0 e + p, \tag{2.9}$$

$$h = \frac{1}{\mu_0} b - m, \tag{2.10}$$

where m is the magnetization, p the polarization, μ_0 the vacuum permeability, and ε_0 the vacuum permittivity. The identity (2.10) will be discussed later in more detail in the context of crustal field modeling.

Many of the geomagnetic phenomena are in change at length scales L of about 100–1000 km and time scales T of hours to days, or even longer, such that the quotient L/T is nearly vanishing in comparison to the speed of light. Under this assumption, the displacement current $\frac{\partial}{\partial t} d$ can be neglected in (2.7). Furthermore, the magnetization m vanishes in the Earth's atmosphere, so that the Eqs. (2.7), (2.8), and (2.10) reduce to the so-called *pre-Maxwell equations*

$$\nabla \wedge b = \mu_0 \, j, \tag{2.11}$$

$$\nabla \cdot b = 0. \tag{2.12}$$

The pre-Maxwell equations form the foundation for external modeling approaches (for the sake of simplicity, some monographs occasionally set the vacuum permeability μ_0 to be equal to 1 in their theoretical part, but the actual value μ_0 must be always observed for numerical purposes). In the neutral atmosphere near the Earth's surface, one can additionally assume that the current density j vanishes, which is a necessary condition for the Gauss representation. At the altitude of low Earth orbiting satellites (approximately 300–600 km), this does not hold true, such that the magnetic field cannot be assumed to be a potential field in these regions. For reasons of content, we will not go into this topic further in our approach. A more detailed description can be found, e.g., in [10].

The *Helmholtz and Mie decomposition* are two decompositions of particular importance in decomposing sources in electromagnetism.

Helmholtz Decomposition The Helmholtz decomposition splits a vector field f into a divergence-free contribution $\nabla \wedge v$ and a curl-free contribution ∇U, such that (see, e.g., [10, 48, 49, 78, 135])

$$f = \nabla U + \nabla \wedge v. \tag{2.13}$$

It can be found in most textbooks related to electromagnetism (e.g., [10, 18]). Mathematically, there exist formulations for several function spaces and underlying domains. In our approach (see also [49, 65]), we formulate the Helmholtz decomposition under the classical condition of continuous differentiability. Nevertheless, it should be pointed out that the assumptions can be weakened in various mathematical ways, which will not be discussed here.

We base our considerations on both the whole Euclidean space $\mathbb{R}^3$ and regular regions $\mathcal{B} \subset \mathbb{R}^3$.

A *regular region* $\mathcal{B}$ is understood to be an open and connected subset in $\mathbb{R}^3$, so that

- $\mathcal{B}$ divides $\mathbb{R}^3$ uniquely into the bounded "inner space" $\mathcal{B}$ and the unbounded "outer space" $\mathcal{B}^c = \mathbb{R}^3 \backslash \overline{\mathcal{B}}$, $\overline{\mathcal{B}} = \mathcal{B} \cup \partial \mathcal{B}$.
- $\partial \mathcal{B}$, i.e., the boundary of $\mathcal{B}$, constitutes an orientable, piecewise smooth Lipschitzian manifold of dimension 2.

Typical examples of regular regions are ball, ellipsoid, spheroid, telluroid, geoid(al potato), (real) Earth's body. Cube, polyhedra, etc. come into play for purposes of exploration.

Helmholtz Decomposition. The following statements are valid:

(a) Let f be of class $c^{(1)}(\mathbb{R}^3)$ satisfying the decay condition

$$|f(x)| = O(|x|^{-(2+\varepsilon)}), \ \varepsilon > 0, \ |x| \to \infty. \tag{2.14}$$

Then there exist functions v of class $c^{(1)}(\mathbb{R}^3)$ and U of class $C^{(1)}(\mathbb{R}^3)$ such that

$$f(x) = \nabla U(x) + \nabla \wedge v(x), \quad x \in \mathbb{R}^3. \tag{2.15}$$

If additionally

$$|\nabla \otimes f(x)| = O(|x|^{-(2+\varepsilon)}), \ \varepsilon > 0, \ |x| \to \infty, \tag{2.16}$$

then the functions U, v can be represented by

$$U(x) = -\frac{1}{4\pi} \int_{\mathbb{R}^3} \frac{\nabla_y \cdot f(y)}{|x - y|} \, dy \tag{2.17}$$

$$= -\frac{1}{4\pi} \int_{\mathbb{R}^3} f(y) \cdot \frac{x - y}{|x - y|^3} \, dy,$$

$$v(x) = \frac{1}{4\pi} \int_{\mathbb{R}^3} \frac{\nabla_y \wedge f(y)}{|x - y|} \, dy \tag{2.18}$$

$$= \frac{1}{4\pi} \int_{\mathbb{R}^3} f(y) \wedge \frac{x - y}{|x - y|^3} \, dy,$$

for $x \in \mathbb{R}^3$.

(b) Let $\mathcal{B} \subset \mathbb{R}^3$ be a regular region with outward directed unit normal field v. Assume that f is of class $c^{(1)}(\mathcal{B})$. Then there exist functions v of class $c^{(1)}(\mathcal{B})$ and U of class $C^{(1)}(\mathcal{B})$ such that

$$f(x) = \nabla U(x) + \nabla \wedge v(x), \quad x \in \mathcal{B}. \tag{2.19}$$

If f is of class $c^{(1)}(\overline{\mathcal{B}})$, then a representation for U, v is given by

$$U(x) = -\frac{1}{4\pi} \int_{\mathcal{B}} \frac{\nabla_y \cdot f(y)}{|x - y|} \, dy + \int_{\partial \mathcal{B}} \frac{v(y) \cdot f(y)}{|x - y|} dS(y) \tag{2.20}$$

$$= -\frac{1}{4\pi} \int_{\mathcal{B}} f(y) \cdot \frac{x - y}{|x - y|^3} \, dy,$$

$$v(x) = \frac{1}{4\pi} \int_{\mathcal{B}} \frac{\nabla_y \wedge f(y)}{|x - y|} \, dy - \int_{\partial \mathcal{B}} \frac{v(y) \wedge f(y)}{|x - y|} \, dS(y) \tag{2.21}$$

$$= \frac{1}{4\pi} \int_{\mathcal{B}} f(y) \wedge \frac{x - y}{|x - y|^3} \, dy,$$

for $x \in \mathcal{B}$.

All proofs can be found in the mathematical monograph [49] on potential theory.

In general, the functions U, v are not uniquely determined since any function $\hat{v} = v + \nabla V$, with V twice continuously differentiable, yields

$$f = \nabla U + \nabla \wedge \hat{v} \tag{2.22}$$

as well. Not even the quantities ∇U and $\nabla \wedge v$ are uniquely determined: For any harmonic function W, there exists a vector field w such that $\nabla W = -\nabla \wedge w$. In consequence,

$$f = \nabla \hat{U} + \nabla \wedge \hat{v}, \tag{2.23}$$

for $\hat{U} = U + W$ and $\hat{v} = v + w$.

Pre-Maxwell Equations Uniqueness can be achieved by adequate decay conditions at infinity or boundary conditions on $\partial \mathcal{B}$. A viable formulation is given by the following statement.

Uniqueness of the Helmholtz Decomposition. The following results hold true:

(a) Let f be of class $c^{(1)}(\mathbb{R}^3)$ with

$$|f(x)| = O(|x|^{-(2+\varepsilon)}), \quad \varepsilon > 0, \quad |x| \to \infty. \tag{2.24}$$

Then the following statements hold true:

(1) *The quantities ∇U and $\nabla \wedge v$ are uniquely determined if U is regular at infinity.*

(2) *The functions U, v are uniquely determined if U is regular at infinity and, additionally,*

$$\nabla \cdot v(x) = 0, \quad x \in \mathbb{R}^3, \tag{2.25}$$

with

$$|v(x)| = O(|x|^{-2}), \quad |x| \to \infty. \tag{2.26}$$

(b) *Let $\mathcal{B} \subset \mathbb{R}^3$ be a regular region, and assume that f is of class $c^{(1)}(\overline{\mathcal{B}})$. Then the following statements are valid:*

(1) *The quantities ∇U and $\nabla \wedge v$ are uniquely determined if $U(x) = 0, x \in \partial \mathcal{B}$.*

(2) *The functions U, v are uniquely determined if $U(x) = 0, x \in \partial \mathcal{B}$, and, additionally,*

$$\nabla \cdot v(x) = 0, \quad x \in \mathcal{B}, \tag{2.27}$$

with v tangential, i.e., $v(x) \cdot v(x) = 0, x \in \partial \mathcal{B}$.

It should be remarked that the special choice of boundary values $U(x) = 0$, $v(x) \cdot v(x) = 0, x \in \partial \mathcal{B}$, as given in part (b) has been made for convenience only. Any other choice of sufficiently smooth boundary functions D_U, D_v, such that $U(x) = D_U(x)$ and $v(x) \cdot v(x) = D_v(x), x \in \partial \mathcal{B}$, also leads to uniqueness.

Concerning a solution of the pre-Maxwell equations, we are not so much interested in determining unique functions U, v for the Helmholtz decomposition but rather in specifying the magnetic field uniquely by its curl and its divergence.

Unique Solvability of the Pre-Maxwell Equations. The following results are valid (see also [65]):

(a) *Let f be of class $c^{(1)}(\mathbb{R}^3)$ satisfying*

$$|f(x)| = O(|x|^{-2}), \quad |x| \to \infty. \tag{2.28}$$

If f additionally satisfies

$$\nabla \wedge f(x) = 0, \quad x \in \mathbb{R}^3, \tag{2.29}$$

$$\nabla \cdot f(x) = 0, \quad x \in \mathbb{R}^3, \tag{2.30}$$

i.e., f is a curl and divergence free, then $f(x) = 0, x \in \mathbb{R}^3$.

(b) *Let $\mathcal{B} \subset \mathbb{R}^3$ be a regular region, and suppose that the vector field f is of class $c^{(1)}(\mathcal{B}) \cap c^{(0)}(\overline{\mathcal{B}})$. If f additionally satisfies*

$$\nabla \wedge f(x) = 0, \quad x \in \mathcal{B}, \tag{2.31}$$

$$\nabla \cdot f(x) = 0, \quad x \in \mathcal{B}, \tag{2.32}$$

$$\nu(x) \cdot f(x) = 0, \quad x \in \partial\mathcal{B}, \tag{2.33}$$

i.e., f is a curl and divergence free tangential field, then $f(x) = 0$, $x \in \mathcal{B}$.

The uniqueness results in connection with the Helmholtz representation lead to the *Law of Biot–Savart* (see, e.g., [49, 116]):
   Let $b \in c^{(2)}(\mathbb{R}^3)$ satisfy the asymptotic relation

$$|b(x)| = O(|x|^{-2}), |x| \to \infty. \tag{2.34}$$

Furthermore, suppose that j is of class $c^{(2)}(\mathbb{R}^3)$ with

$$|j(x)| = O(|x|^{-(2+\varepsilon)}), \ \varepsilon > 0, \ |x| \to \infty. \tag{2.35}$$

and

$$|\nabla \otimes j(x)| = O(|x|^{-(2+\varepsilon)}), \ \varepsilon > 0, \ |x| \to \infty. \tag{2.36}$$

If the pre-Maxwell equations

$$\nabla \wedge b(x) = \mu_0 \, j(x), \quad x \in \mathbb{R}^3, \tag{2.37}$$

$$\nabla \cdot b(x) = 0, \quad\quad x \in \mathbb{R}^3, \tag{2.38}$$

are satisfied, then b admits an integral representation of the form

$$b(x) = \frac{\mu_0}{4\pi} \int_{\mathbb{R}^3} j(y) \wedge \frac{x-y}{|x-y|^3} \, dy, \quad x \in \mathbb{R}^3. \tag{2.39}$$

A local version of the law of Biot–Savart on regular regions $\mathcal{B} \subset \mathbb{R}^3$ can be formulated as well. Clearly, adequate boundary values on $\partial\mathcal{B}$ have to be required to obtain uniqueness of b.

The decay conditions in the mathematical context above are rather strict. It is a well-known fact that the uniform decay to zero at infinity is sufficient for many assertions. We have chosen the stricter conditions to guarantee the existence of the integral representations. Furthermore, most geoscientifically relevant magnetic fields satisfy these decay conditions (e.g., a dipole field behaves like $|x|^{-3}$ at infinity), and the current systems j are typically restricted to bounded regions like the ionosphere.

Mie Decomposition Next, we turn to the second important decomposition in geomagnetic modeling, namely, the Mie decomposition. We start with the following definitions (cf. [6]):

Let $\mathcal{B} \subset \mathbb{R}^3$ be a regular region with outward directed unit normal field ν.

Solenoidal Field   A vector field f of class $c^{(1)}(\mathcal{B})$ is said to be *solenoidal* if

$$\int_{\partial \mathcal{G}} \nu(y) \cdot f(y) \, dS(y) = 0 \tag{2.40}$$

for every boundary $\partial \mathcal{G}$ of a regular region $\mathcal{G} \subset \mathcal{B}$ with $\mathrm{dist}(\partial \mathcal{G}, \partial \mathcal{B}) > 0$.

Toroidal Field   A vector field f of class $c^{(1)}(\mathcal{B})$ is called *toroidal* if a scalar field Q of class $C^{(1)}(\mathcal{B})$ (or at least sufficiently smooth to allow the application of the curl gradient L) exists such that

$$f(x) = LQ(x) = x \wedge \nabla Q(x), \quad x \in \mathcal{B}. \tag{2.41}$$

Poloidal Field   A vector field f of class $c^{(1)}(\mathcal{B})$ is called *poloidal* if a scalar field P of class $C^{(2)}(\mathcal{B})$ (or at least sufficiently smooth to allow the application of the operator $\nabla \wedge L$) exists such that

$$f(x) = \nabla \wedge LP(x), \quad x \in \mathcal{B}. \tag{2.42}$$

The fields P and Q are also known as *Mie scalars*.

As a consequence of the theorem of Gauss, any solenoidal vector field is divergence-free (i.e., $\nabla \cdot f = 0$). Concerning the entire space $\mathbb{R}^3$, the converse holds true as well. Thus, functions satisfying the pre-Maxwell equations everywhere are solenoidal. In general regions $\mathcal{B} \subset \mathbb{R}^3$, however, a divergence-free function is not necessarily solenoidal.

The next statements are formulated in spherical shells ("ball rings")

$$\mathbb{B}^3_{R_0, R_1}(x) = \mathbb{B}^3_{R_1}(x) \backslash \overline{\mathbb{B}^3_{R_0}(x)} \tag{2.43}$$

in $\mathbb{R}^3$, i.e., the difference between the open ball $\mathbb{B}^3_{R_1}(x)$ of radius R_1 around x and the closed ball $\overline{\mathbb{B}^3_{R_0}(x)}$ of radius $R_0 < R_1$ around x :

Mie Decomposition. Let $f : \mathbb{B}^3_{R_0, R_1} \to \mathbb{R}^3$ be a solenoidal vector field. Then there exist scalar fields P, Q of class $C^{(1)}(\mathbb{B}^3_{R_0, R_1})$ such that

$$f(x) = \nabla \wedge LP(x) + LQ(x), \quad x \in \mathbb{B}^3_{R_0, R_1}. \tag{2.44}$$

P, Q are determined uniquely by the additional conditions

$$\frac{1}{4\pi R^2} \int_{\mathbb{S}^2_R} P(y) \, dS(y) = \frac{1}{4\pi R^2} \int_{\mathbb{S}^2_R} Q(y) \, dS(y) = 0, \tag{2.45}$$

for every $R \in (R_0, R_1)$.

The proof requires techniques intrinsic to the sphere, as treated, e.g., in [16, 49]. For the Mie scalars P, Q we see that they are only determined up to a constant (due to the assumption of vanishing integral mean values). However, the poloidal part $p = \nabla \wedge LP$ and the toroidal part $q = LQ$ are determined uniquely without further assumptions.

One of the most significant properties of the Mie decomposition is its capability to reduce the vectorial pre-Maxwell equations (2.11) and (2.12) to a set of simple scalar equations. More precisely, let P_b, Q_b be the *Mie scalars* of the magnetic induction b, and P_j, Q_j the Mie scalars of the current density j. Then

$$\nabla \wedge b = \nabla \wedge (\nabla \wedge LP_b + LQ_b) \tag{2.46}$$
$$= \nabla(\nabla \cdot LP_b) - \Delta LP_b + \nabla \wedge LQ_b$$
$$= -L\Delta P_b + \nabla \wedge LQ_b,$$

where we have used $\nabla \cdot L = 0$ and $\Delta L = L\Delta$ in the last step. Since $\nabla \wedge b = j$ and $j = \nabla \wedge LP_j + LQ_j$, the uniqueness of the Mie scalars (under the assumption of vanishing integral mean values) yields

$$Q_b = P_j, \tag{2.47}$$
$$\Delta P_b = -Q_j \tag{2.48}$$

(note that the vacuum permeability μ_0 is chosen here to be equal to one). Thus, the original pre-Maxwell equations have been reduced to scalar equations involving the Laplace operator, which can be handled by potential-theoretic tools.

It should be mentioned that Eqs. (2.47) and (2.48) show that toroidal magnetic fields are solely generated by poloidal currents, and poloidal magnetic fields are solely generated by toroidal currents.

Because of its significance, the previous result is summarized as follows:

Let $b \in c^{(2)}(\mathbb{R}^3)$ and $j \in c^{(1)}(\mathbb{R}^3)$ satisfy the pre-Maxwell equations

$$\nabla \wedge b(x) = j(x), \quad x \in \mathbb{R}^3, \tag{2.49}$$
$$\nabla \cdot b(x) = 0, \quad x \in \mathbb{R}^3. \tag{2.50}$$

Furthermore, let P_b, Q_b, P_j, Q_j be the uniquely determined Mie scalars of b and j, respectively. Then

$$Q_b(x) = P_j(x), \quad x \in \mathbb{R}^3, \tag{2.51}$$
$$\Delta P_b(x) = -Q_j(x), \quad x \in \mathbb{R}^3. \tag{2.52}$$

We conclude this section with some properties of poloidal and toroidal fields (see, e.g., [6, 10, 49]):

Let $f : \mathbb{B}^3_{R_0, R_1} \to \mathbb{R}^3$, $0 \leq R_0 \leq R \leq R_1$, $R_0 < R_1$, be a sufficiently often differentiable vector field. Then the following statements are valid:

- f is toroidal if and only if f is solenoidal and tangential.
- f is poloidal if and only if f is solenoidal and $\nabla \wedge f$ is tangential.
- If f is toroidal, then $\nabla \wedge f$ is poloidal. And vice versa, if f is poloidal, then $\nabla \wedge f$ is toroidal.

2.4 Gauss Representation

In source-free regions with vanishing current density j, we have

$$\nabla \wedge b = 0, \tag{2.53}$$

$$\nabla \cdot b = 0, \tag{2.54}$$

and the magnetic field b can be represented as a potential field, i.e.,

$$b = \nabla B, \tag{2.55}$$

where B is a harmonic function. For the geomagnetic field, the assumption of a vanishing current density holds true only in the neutral atmosphere, i.e., the region between the Earth's surface and the ionosphere. In our spherical approximation of the Earth, this means that the identity (2.55) holds true in the spherical shell $\mathbb{B}^3_{R_0, R_1}$, with R_0 denoting a mean Earth radius and R_1 a radius up to the lower bound of the ionosphere. Since U is harmonic, we can use (spherical harmonics) methods of classical potential theory for an approximation of the magnetic field.

We have already mentioned the difficulty of separating the different contributions to the Earth's magnetic field. It is, however, possible to at least separate the contributions due to sources in the exterior of a given sphere $\mathbb{S}^2_R$ (or a spherical shell $\mathbb{B}^3_{R_0, R_1}$) from those due to sources in the interior. This is of interest, e.g., when modeling the Earth's interior magnetic field. As a consequence, exterior influences like magnetospheric currents can be filtered out using this separation.

In the case that the magnetic measurements are conducted in the source-free "ball ring" $\mathbb{B}^3_{R_0, R_1}$, the *Gauss representation* provides such a separation (for an illustration, see Fig. 2.7). More precisely (see, e.g., [50]), the magnetic field

$$b = \nabla B \tag{2.56}$$

can be split into

$$b(x) = b^{\text{int}}(x) + b^{\text{ext}}(x), \quad x \in \mathbb{B}^3_{R_0, R_1}, \tag{2.57}$$

with

$$\nabla \wedge b^{\text{int}}(x) = 0, \quad x \in \mathbb{R}^3 \setminus \overline{\mathbb{B}^3_{R_0}}, \tag{2.58}$$

$$\nabla \cdot b^{\text{int}}(x) = 0, \quad x \in \mathbb{R}^3 \setminus \overline{\mathbb{B}^3_{R_0}}, \tag{2.59}$$

and

$$\nabla \wedge b^{\text{ext}}(x) = 0, \quad x \in \mathbb{B}^3_{R_1}, \tag{2.60}$$

$$\nabla \cdot b^{\text{ext}}(x) = 0, \quad x \in \mathbb{B}^3_{R_1}. \tag{2.61}$$

Thus, b^{int} denotes the part of the magnetic field due to sources inside the Earth $\mathbb{B}^3_{R_0}$ and can be represented as a potential field $b^{\text{int}}(x) = \nabla B^{\text{int}}(x)$, $x \in \mathbb{R}^3 \setminus \overline{\mathbb{B}^3_{R_0}}$, where

$$B^{\text{int}}(x) = \sum_{n=0}^{\infty} \sum_{k=1}^{2n+1} \beta_{n,k} H^{R_0}_{-n-1,k}(x), \quad x \in \mathbb{R}^3 \setminus \overline{\mathbb{B}^3_{R_0}}, \tag{2.62}$$

and $\{H^{R_0}_{-n-1,k}\}$ is a system of outer harmonics given by

$$H^{R_0}_{-n-1,k}(x) = \frac{1}{4\pi R_0} \left(\frac{R_0}{|x|} \right)^{n+1} Y_{n,k}\left(\frac{x}{|x|} \right). \tag{2.63}$$

Analogously, b^{ext} denotes the part of the magnetic field due to sources in the iono- and magnetosphere, i.e., in the exterior $\mathbb{R}^3 \setminus \overline{\mathbb{B}^3_{R_1}}$, and can be represented as a potential field $b^{\text{ext}}(x) = \nabla B^{\text{ext}}(x)$, $x \in \mathbb{B}^3_{R_1}$, where

$$B^{\text{ext}}(x) = \sum_{n=0}^{\infty} \sum_{k=1}^{2n+1} \alpha_{n,k} H^{R_1}_{n,k}(x), \quad x \in \mathbb{B}^3_{R_1}, \tag{2.64}$$

and $\{H^{R_1}_{n,k}\}$ is a system of inner harmonics given by

$$H^{R_1}_{n,k}(x) = \frac{1}{4\pi R_1} \left(\frac{|x|}{R_1} \right)^{n} Y_{n,k}\left(\frac{x}{|x|} \right). \tag{2.65}$$

The coefficients $\alpha_{n,k}$ and $\beta_{n,k}$ may be understood as Fourier expansion coefficients in the conventional sense of spherical harmonic theory.

However, if the measurements are not conducted in a source-free region (i.e., we have a current density $j \neq 0$), as is the case for satellite missions (say, orbiting on a sphere $\mathbb{S}^2_R$ of radius $R > R_1$), the magnetic field b cannot be represented by a potential field ∇B as in the case of the Gauss representation. As a remedy, the *Mie*

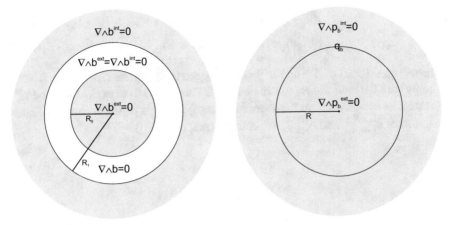

Fig. 2.7 Separation into interior and exterior sources: with respect to measurements in a source-free shell (left) and measurements at satellite altitude (right)

representation is used. If $p_b = \nabla \wedge L P_b$ and $q_b = L Q_b$ denote the poloidal and toroidal part of the magnetic field and $p_j = \nabla \wedge L P_j$ and $q_j = L Q_j$ the poloidal and toroidal part of the current density, then the magnetic field can be split into

$$b(x) = p_b^{\text{int}}(x) + p_b^{\text{ext}}(x) + q_b(x), \quad x \in \mathbb{S}_R^2, \tag{2.66}$$

with

$$\nabla \wedge p_b^{\text{int}}(x) = \begin{cases} 0, & x \in \mathbb{R}^3 \setminus \overline{\mathbb{B}_R^3}, \\ q_j(x), & x \in \mathbb{B}_R^3, \end{cases} \tag{2.67}$$

$$\nabla \cdot p_b^{\text{int}}(x) = 0, \quad x \in \mathbb{R}^3, \tag{2.68}$$

and

$$\nabla \wedge p_b^{\text{ext}}(x) = \begin{cases} q_j(x), & x \in \mathbb{R}^3 \setminus \overline{\mathbb{B}_R^3}, \\ 0, & x \in \mathbb{B}_R^3, \end{cases} \tag{2.69}$$

$$\nabla \cdot p_b^{\text{ext}}(x) = 0, \quad x \in \mathbb{R}^3 \tag{2.70}$$

(for an illustration, see again Fig. 2.7). The toroidal magnetic field q_b is closely related to the radial current density at the satellite's orbit $\mathbb{S}_R^2$. It represents the part of the magnetic field that is due to poloidal currents crossing the orbital sphere. The poloidal part p_b can be split into a part p_b^{int} due to toroidal currents in the interior of the satellite's orbit, i.e., $\mathbb{B}_R^3$, and a part p_b^{ext} due to toroidal currents in the exterior, i.e., $\mathbb{R}^3 \setminus \overline{\mathbb{B}_R^3}$. Just as for the source-free case in (2.57)–(2.64), we find potentials B^{int}, B^{ext} such that

$$p_b^{\text{int}}(x) = \nabla B^{\text{int}}(x), \quad x \in \mathbb{R}^3 \setminus \overline{\mathbb{B}_R^3}, \tag{2.71}$$

$$p_b^{\text{ext}}(x) = \nabla B^{\text{ext}}(x), \quad x \in \mathbb{B}_R^3. \tag{2.72}$$

The representations of B^{int} and B^{ext} are analogous to (2.62) and (2.64), respectively, simply with the radii R_0, R_1 substituted by R. This allows us to expand p_b^{int} and p_b^{ext} (as well as b^{int} and b^{ext} in the source-free case) in terms of the vector outer/inner harmonics $h_{-n-1,k}^{(1);R}$ and $h_{n,k}^{(2);R}$, respectively (see, e.g., [49]).

Chapter 3
Dipole Potential Based Magnetometry

Magnetometry, being another potential method, has a lot in common with gravimetry and, hence, they are often measured and interpreted together. A decorrelation methodology similar to that of gravimetry is also feasible. However, there are also significant differences. Once again, there are no magnetic monopoles (in contrast to gravity) and, hence, dipoles are principal constituents. The magnetic field of the Earth is less stable than the gravity field. It is changing quickly. In contrast to gravity maps, the magnetic maps are dominated more essentially by local anomalies. Magnetic poles can be repulsive or attractive; they not always attractive as in the gravity case. The differences in magnetization of different rock types are often quite large (much larger differences than in the case of mass density distributions).

3.1 Magnetometric Measurements

Magnetic surveying is the oldest method of geophysical prospecting, but it has become relegated to a method of minor importance because of the advent of seismic reflection surveying in the last few decades.

The magnetic anomalies can originate from a series of changes in lithology, variations in the magnetized bodies thickness, faulting, pleats, and topographic relief. The main objective of magnetometry in geoexploration is to contribute with information about the relationship among the thermal activity, the tectonic and stratigraphy of the area by means of the interpretation of the underground magnetic properties. In this respect, it should be mentioned that most of the rocks are not magnetic. However, certain types of rocks contain enough minerals to originate significant magnetic anomalies. The data interpretation that reflects differences in local abundance of magnetization is especially useful to locate faults.

Survey magnetometers can be roughly divided into two basic types: Scalar magnetometers measure the total strength of the magnetic field to which they

C. Blick et al., *Inverse Magnetometry*, Lecture Notes in Geosystems Mathematics and Computing, https://doi.org/10.1007/978-3-030-79508-5_3

are subjected, but not its direction. Vector magnetometers have the capability to measure the component of the magnetic field in a particular direction, relative to the spatial orientation of the device. Currently the most often used types are the proton-precession (Overhauser magnetometer), fluxgate, and optically pumped magnetometers. The proton and fluxgate magnetometers have a similar sensitivity (about $1 - 0.1\,\mathrm{nT}$), the difference is in the components of the magnetic flux measured. The proton magnetometer measures the total field, the overall amplitude of the magnetic field, whereas the fluxgate magnetometers measure individual components. The fluxgate magnetometers are capable of continuous measurements and, hence, are used for airborne, ship, and satellite measurements. In contrast, the proton magnetometers do not have a drift and are common in ground surveys. The optically pumped magnetometers offer much higher sensitivity ($0.001\,\mathrm{nT}$) and also high frequency of readings (up to $1000\,\mathrm{Hz}$). They are often used for archaeological prospection (for more details the reader is referred, e.g., to the lecture notes [139] and the references therein).

Measurements of magnetic intensities are usually easier to carry out than gravimetric ones, and the measurements are very quick, hence, large areas are easily covered.

Common uses of magnetometers are as follows (see also the list and comments given by Rivas [131]):

• Mineral and geothermal exploration,
• Fault studies,
• Locating buried utilities (tanks and drums, pipelines, etc.),
• Archaeological research.

3.2   Localized Currents

Next we deal with the characteristics of the crustal field, which are of interest in geoexploration. For this purpose, we assume that *all currents are located inside* $\overline{\mathcal{B}}$, where $\mathcal{B} \subset \mathbb{R}^3$ is a regular region. The law of Biot–Savart then leads to the representation

$$
\begin{aligned}
b(x) &= \frac{\mu_0}{4\pi} \int_{\mathcal{B}} j(y) \wedge \frac{x-y}{|x-y|^3}\, dy \\
&= \frac{\mu_0}{4\pi} \int_{\mathcal{B}} j(y) \wedge \left(\nabla_y \frac{1}{|x-y|} \right) dy \\
&= -\frac{\mu_0}{4\pi} \int_{\mathcal{B}} j(y) \wedge \left(\nabla_x \frac{1}{|x-y|} \right) dy.
\end{aligned}
\tag{3.73}
$$

Now, by virtue of the product rule, we have

$$j(y) \wedge \left(\nabla_x \frac{1}{|x-y|} \right) = \underbrace{\left(\frac{1}{|x-y|} \right) \nabla_x \wedge j(y)}_{=0} - \nabla_x \wedge \left(\frac{j(y)}{|x-y|} \right). \tag{3.74}$$

This shows that

$$b(x) = \frac{\mu_0}{4\pi} \int_{\mathcal{B}} j(y) \wedge \frac{x-y}{|x-y|^3} \, dy \tag{3.75}$$

$$= \nabla \wedge a(x),$$

for $x \in \mathbb{R}^3$, where a is the vectorial Newton volume potential

$$a(x) = \frac{\mu_0}{4\pi} \int_{\mathcal{B}} \frac{j(y)}{|x-y|} \, dy = \mu_0 \int_{\mathcal{B}} G(\Delta; |x-y|) \, j(y) \, dy. \tag{3.76}$$

$G(\Delta; |\cdot - \cdot|)$ is the *Newton kernel,* i.e., the fundamental solution of the negative Laplace operator $-\Delta$ given by

$$G(\Delta; |x-y|) = \frac{1}{4\pi} \frac{1}{|x-y|}, \quad x, y \in \mathbb{R}^3, \ x \neq y. \tag{3.77}$$

It follows that

$$\nabla \wedge b(x) = \nabla \wedge (\nabla \wedge a(x)) = \nabla(\nabla \cdot a(x)) - \Delta a(x), \quad x \in \mathbb{R}^3. \tag{3.78}$$

Moreover, it is known that

$$\Delta a(x) = \begin{cases} -\mu_0 j(x), & x \in \mathcal{B}, \\ 0, & x \in \mathbb{R}^3 \setminus \mathcal{B}, \end{cases} \tag{3.79}$$

which leads to

$$\nabla \wedge b(x) = \begin{cases} \mu_0 j(x) + \nabla(\nabla \cdot a(x)), & x \in \mathcal{B}, \\ \nabla(\nabla \cdot a(x)), & x \in \mathbb{R}^3 \setminus \mathcal{B}. \end{cases} \tag{3.80}$$

We see that the representation of b as chosen in (3.75) includes additional sources $\nabla(\nabla \cdot a)$. The term $\nabla \cdot a$ can be rewritten in the form

$$\nabla_x \cdot a(x) = -\frac{\mu_0}{4\pi} \int_{\mathcal{B}} \nabla_y \cdot \left(\frac{j(y)}{|x-y|} \right) dV(y) + \underbrace{\frac{\mu_0}{4\pi} \int_{\mathcal{B}} \frac{\nabla_y \cdot j(y)}{|x-y|} \, dy}_{=0} \tag{3.81}$$

$$= -\frac{\mu_0}{4\pi} \int_{\partial \mathcal{B}} \frac{v(y) \cdot j(y)}{|x-y|} \, dS(y),$$

where $\nabla \cdot j = 0$ and the theorem of Gauss have been used (since our considerations here have only a motivating character, all appearing functions are supposed to be sufficiently smooth and have a sufficient decay at infinity). $\nabla(\nabla \cdot a)$ can be interpreted as an electric field generated by a surface charge density due to the normal current density $v \cdot j$ on ∂B.

Heuristic Motivation of the Dipole Magnetic Induction We go back to the representation (3.75) for b localized in a regular region B

$$b(x) = \nabla \wedge a(x), \tag{3.82}$$

where a is the vectorial Newton integral

$$a(x) = \frac{\mu_0}{4\pi} \int_B \frac{j(y)}{|x - y|} \, dy, \quad x \in \mathbb{R}^3. \tag{3.83}$$

Our purpose is to motivate the representation of the dipole magnetic induction heuristically: We formally replace the magnetic vector potential a in (3.82) by a "multipole expansion"

$$a(x) = \frac{\mu_0}{4\pi} \int_B \frac{j(y)}{|x - y|} \, dy \tag{3.84}$$

$$= \frac{\mu_0}{4\pi} \left(\frac{1}{|x|} \int_B j(y) \, dy + \frac{1}{|x|^3} \int_B j(y)(x \cdot y) \, dy + \dots \right),$$

which is at least valid in a far-field region. Since j is localized, it is not difficult to see that the first term on the right side of (3.84), i.e., the monopole term vanishes (as the volume enclosing currents is limited, but the boundary surface may be placed far away from the currents, so that the regularity condition comes into play):

$$\int_B j(y) \, dy = 0. \tag{3.85}$$

An elementary calculation yields the identity

$$\int_B j(y)(x \cdot y) \, dy = -\frac{1}{2} x \wedge \int_B y \wedge j(y) \, dy. \tag{3.86}$$

Convention: The dipole term (3.86) is dominant in the expansion (3.84). Thus, in the literature, the vector potential a in (3.84) is given by

$$a(x) = \frac{\mu_0}{4\pi} \int_B \frac{j(y)}{|x - y|} \, dy \simeq \frac{\mu_0}{4\pi} \frac{m_0 \wedge x}{|x|^3}, \tag{3.87}$$

where the *magnetic moment* m_0 is defined by

$$m_0 = \frac{1}{2} \int_{\mathcal{B}} y \wedge j(y)\, dy = \int_{\mathcal{B}} m(y)\, dy, \tag{3.88}$$

where the last term in (3.88) is the integral over the *magnetization* (dipole magnetic moment per unit volume) with

$$m(y) = \frac{1}{2} (y \wedge j(y)). \tag{3.89}$$

The *magnetic induction,* more accurately, *the dipole magnetic induction b* follows from the curl of the right side of (3.87) so that

$$b(x) \cong \nabla \wedge \frac{\mu_0}{4\pi} \left(m_0 \wedge \frac{x}{|x|^3} \right). \tag{3.90}$$

An easy calculation yields

$$b(x) \cong \frac{\mu_0}{4\pi} \left(m_0 \left(\nabla \cdot \frac{x}{|x|^3} \right) - \frac{x}{|x|^3} (\nabla \cdot m_0) + \left(\frac{x}{|x|^3} \cdot \nabla \right) m_0 - (m_0 \cdot \nabla) \frac{x}{|x|^3} \right). \tag{3.91}$$

Since m_0 is independent of x and $\nabla \cdot \frac{x}{|x|^3} = 0$ for $|x| \neq 0$, we obtain

$$b(x) \cong -\frac{\mu_0}{4\pi} (m_0 \cdot \nabla) \frac{x}{|x|^3} = -\frac{\mu_0}{4\pi} (m_0 \cdot \nabla) \nabla \frac{1}{|x|} = \frac{\mu_0}{4\pi} \frac{3(m_0 \cdot \frac{x}{|x|}) \frac{x}{|x|} - m_0}{|x|^3}, \tag{3.92}$$

hence,

$$b(x) \cong \frac{\mu_0}{4\pi} \frac{3(m_0 \cdot x)x}{|x|^5} - \frac{\mu_0}{4\pi} \frac{m_0}{|x|^3}. \tag{3.93}$$

The right side of (3.93) yields the representation of the dipole magnetic induction.

3.3   Vector Potential Representation Involving Magnetization

We assume that matter acquires a macroscopic magnetic moment distribution (note that, by macroscopic, it is meant averaged over a large number of atoms and, under this assumption, the concept of a distribution characterized by the magnetic moment per unit volume, i.e., magnetization m, is reasonable).

In accordance with (3.87), we are (approximately) led to start our next considerations from the infinitesimal relation

$$d(a(y)) = \frac{\mu_0}{4\pi} m(y) \wedge \frac{x - y}{|x - y|^3}\, dy, \tag{3.94}$$

so that a may be written as follows:

$$a(x) = \frac{\mu_0}{4\pi} \int_B m(y) \wedge \frac{x - y}{|x - y|^3} \, dy. \tag{3.95}$$

Equivalently we have

$$a(x) = \frac{\mu_0}{4\pi} \int_B m(y) \wedge \nabla_y \frac{1}{|x - y|} \, dy. \tag{3.96}$$

The right side of (3.96) can be rewritten in the form

$$\frac{\mu_0}{4\pi} \int_B m(y) \wedge \nabla_y \frac{1}{|x - y|} \, dy = -\frac{\mu_0}{4\pi} \int_B \nabla_y \wedge \frac{m(y)}{|x - y|} \, dy \tag{3.97}$$

$$+ \frac{\mu_0}{4\pi} \int_B \frac{\nabla_y \wedge m(y)}{|x - y|} \, dy.$$

The first integral on the right side of (3.97) may be expressed as surface integral

$$\frac{\mu_0}{4\pi} \int_B \nabla_y \wedge \frac{m(y)}{|x - y|} \, dy = \frac{\mu_0}{4\pi} \int_{\partial B} \frac{\nu(y) \wedge m(y)}{|x - y|} \, dS(y). \tag{3.98}$$

In fact, by standard arguments (see, e.g., [115]), it follows that the surface integral vanishes.

Summarizing our considerations, we are therefore led to conclude that

$$a(x) = \frac{\mu_0}{4\pi} \int_B \frac{\nabla_y \wedge m(y)}{|x - y|} \, dy. \tag{3.99}$$

It follows that

$$b(x) = \nabla_x \wedge a(x) = \nabla_x \wedge \frac{\mu_0}{4\pi} \int_B \frac{j(y)}{|x - y|} \, dy \tag{3.100}$$

is given by

$$b(x) = \nabla_x \wedge \frac{\mu_0}{4\pi} \int_B \frac{\nabla_y \wedge m(y)}{|x - y|} \, dy.$$

From (3.96) we see that

$$b(x) = \frac{\mu_0}{4\pi} \int_B \nabla_x \wedge \left(m(y) \wedge \frac{x - y}{|x - y|^3} \right) dy \tag{3.101}$$

$$= \frac{\mu_0}{4\pi} \int_B \left(m(y) \left(\nabla_x \cdot \frac{x - y}{|x - y|^3} \right) - \frac{x - y}{|x - y|^3} (\nabla_x \cdot m(y)) \right) dy$$

$$+\frac{\mu_0}{4\pi} \int_B \left(\left(\frac{x-y}{|x-y|^3} \cdot \nabla_x \right) m(y) - (m(y) \cdot \nabla_x) \frac{x-y}{|x-y|^3} \right) dy.$$

By the same arguments applied to (3.91), we obtain the *magnetic dipole induction* in the form

$$b(x) = \nabla_x \frac{\mu_0}{4\pi} \int_B \frac{m(y) \cdot (x-y)}{|x-y|^3} \, dy \qquad (3.102)$$

$$= \nabla_x \frac{\mu_0}{4\pi} \int_B m(y) \cdot \nabla_y \frac{1}{|x-y|} \, dy$$

$$= \nabla_x B(x),$$

where B is the *dipole potential*

$$B(x) = \frac{\mu_0}{4\pi} \int_B m(y) \cdot \nabla_y \frac{1}{|x-y|} \, dy. \qquad (3.103)$$

3.4   Dipole Potentials

Before we continue with the discussion of the macroscopic equations of the geomagnetic theory and their consequences to crustal field modeling, we make a mathematical excursion into the concepts of dipole integrals in Euclidean space $\mathbb{R}^3$.

A magnetic dipole is the simplest observed form, in which magnetism occurs (see, e.g., [10, 18, 28, 34, 73, 82, 85, 115, 116, 129, 130], and the contributions in [68]). In fact, all magnetic structures in our approach to crustal field modeling can be made up of dipoles. Although there are speculations about magnetic monopoles, none has been discovered so far. Magnetic multipoles (quadrupoles and multipoles of higher order) have to be considered in principle, but the far magnetic field essentially approximates the dipole field.

To illustrate the model of the dipole (see, e.g., [18] for more details), we assume an arrangement of two opposing but equally sized point charges in the form of two monopoles (cf. Fig. 3.8, left). A dipole is now defined in that the distance between the two point masses tends toward zero so that the dipole moment remains constant (and thus finite). The dipole defined in this way then lies in a fixed point in space (cf. Fig. 3.8, right).

Point Potential of a Dipole   The dipole potential at a point y is given by

$$B(x) = \frac{\mu_0}{4\pi} m_y \cdot \nabla_y \frac{1}{|x-y|} \qquad (3.104)$$

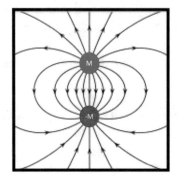 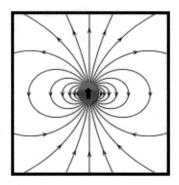

Fig. 3.8 Display of a resulting dipole with field lines of two opposite point masses

$$= -\frac{\mu_0}{4\pi} m_y \cdot \nabla_x \frac{1}{|x - y|}$$

$$= -\frac{\mu_0}{4\pi} \frac{m_y \cdot (x - y)}{|x - y|^3}, \quad m \in \mathbb{R}^3, \ x \in \mathbb{R}^3 \backslash \{y\}.$$

Hence, the dipole potential is generated by differentiating a monopole in the direction of a vector m.

The *dipole field b* generated by m_y is defined (cf. (1.1)) by

$$b(x) = \nabla_x \frac{\mu_0}{4\pi} m_y \cdot \nabla_y \frac{1}{|x - y|} \qquad (3.105)$$

$$= -\nabla_x \frac{\mu_0}{4\pi} \frac{m_y \cdot (x - y)}{|x - y|^3}$$

$$= \frac{\mu_0}{4\pi} \left(3 \, m_y \cdot (x - y) \frac{(x - y)}{|x - y|^5} - \frac{m}{|x - y|^3} \right),$$

$m \in \mathbb{R}^3, \ x \in \mathbb{R}^3 \backslash \{y\}$. It is evident that

$$\nabla \cdot b(x) = \nabla \cdot \nabla B(x) = \Delta B(x) = 0, \quad x \in \mathbb{R}^3 \backslash \{y\}, \qquad (3.106)$$

that is, the dipole field b is divergence-free.

Due to the absence of magnetic monopoles, magnetic fields may be always composed of magnetic dipoles and their superpositions.

Finite Dipole Sum Field in the Directions of $m_{y_i}, i = 1, \ldots, N$ The dipole potential for x in the N directions $m_{y_i}, i = 1, \ldots, N$ consists of the sum of the individual contributions

$$B(x) = \frac{\mu_0}{4\pi} \sum_{i=1}^{N} m_{y_i} \cdot \nabla_y \frac{1}{|x-y|}\bigg|_{y=y_i} \tag{3.107}$$

$$= -\frac{\mu_0}{4\pi} \sum_{i=1}^{N} \frac{m_{y_i} \cdot (x-y_i)}{|x-y_i|^3}, \quad m_{y_i} \in \mathbb{R}^3, \quad x \in \mathbb{R}^3 \backslash \{y_1, \ldots, y_N\}.$$

It is clear that

$$b(x) = \nabla_x \frac{\mu_0}{4\pi} \sum_{i=1}^{N} m_{y_i} \cdot \nabla_y \frac{1}{|x-y|}\bigg|_{y=y_i} \tag{3.108}$$

$$= \frac{\mu_0}{4\pi} \sum_{i=1}^{N} \left(3\, m_{y_i} \cdot (x-y_i) \frac{(x-y_i)}{|x-y_i|^5} - \frac{m_{y_i}}{|x-y_i|^3} \right).$$

Moreover, b is divergence-free outside $\{y_1, \ldots, y_n\}$, i.e., we have

$$\nabla \cdot b(x) = \nabla \cdot \nabla B(x) = \Delta B(x) = 0, \quad x \in \mathbb{R}^3 \backslash \{y_1, \ldots, y_N\}. \tag{3.109}$$

Surface Potential of a Double Layer A double layer on a surface ∂B can be understood as two single assignments separated by an (infinitesimally) small distance σ. The surface normal ν in $y \in \partial B$ intersects the two assignments in points $y + \frac{\sigma}{2}\nu(y)$ and $y - \frac{\sigma}{2}\nu(y)$ (for small values σ), which have area densities of equal size and opposite sign. Each corresponding pair $(y - \frac{\sigma}{2}\nu(y), y + \frac{\sigma}{2}\nu(y))$ of points defines by the Taylor formula (in linearized sense for $\sigma \to 0$) a *potential of a dipole*

$$\frac{\mu_0}{4\pi} M \,\nu(y) \cdot \nabla_y \frac{1}{|x-y|}, \quad x \in \mathbb{R}^3 \backslash \{y\} \tag{3.110}$$

with the dipole moment M. If we integrate over all dipoles to ∂B, then we are led to the *potential of a double layer* (see e.g., [49, 142] for more details):

$$B(x) = \frac{\mu_0}{4\pi} \int_{\partial B} \nu(y) \cdot \nabla_y \frac{1}{|x-y|} F(y)\, dS(y), \quad F \in L^2(\partial B). \tag{3.111}$$

B is harmonic in the "interior" B and in the "exterior" $B^c = \mathbb{R}^3 \backslash \overline{B}$, that is

$$\Delta B(x) = 0, \quad x \notin \partial B. \tag{3.112}$$

At infinity, the surface potential of a double layer behaves like the gradient of the dipole point potential

$$B(x) = O(|x|^{-2}), \quad |x| \to \infty. \tag{3.113}$$

The double layer potential (3.110) must be sharply distinguished from the single layer potential (see, e.g., [87]). Common to both is the fact that they vanish at infinity and satisfy the Laplace equation in the inner and outer space of ∂B. On the surface ∂B itself, however, they have a completely different nature, and it is this difference that makes these fictitious potentials mathematically useful. Indeed, layer potentials help us to handle inner and outer boundary value problems by surface integral equations over ∂B. The essential tool for their solution is the Fredholm theory (see, e.g., [49, 102, 142]).

Magnetostatic Volume Potential from Magnetization Distribution Let $B \subset \mathbb{R}^3$ be a regular region representing the test area for exploration. Suppose that the dipole densities are distributed continuously over $B \subset \mathbb{R}^3$, such that $m : x \mapsto m(x), x \in \overline{B}$, i.e., the spatial distribution of dipole densities is the so-called *magnetization*. Then, the discrete sum (3.107) becomes a continuous sum, i.e., an integral over the body B, so that

$$
\begin{aligned}
B(x) &= \frac{\mu_0}{4\pi} \int_B m(y) \cdot \nabla_y \frac{1}{|x-y|} \, dy & (3.114) \\
&= -\frac{\mu_0}{4\pi} \int_B \frac{m(y) \cdot (x-y)}{|x-y|^3} \, dy \\
&= \frac{\mu_0}{4\pi} \int_B \nabla_x \cdot \left(-\frac{1}{|x-y|} m(y) \right) \, dy \\
&= -\nabla_x \cdot \frac{\mu_0}{4\pi} \int_B m(y) \frac{1}{|x-y|} \, dy,
\end{aligned}
$$

where, as usual, dy is the volume element with respect to the variable y.

B is the *dipole potential of a distribution of magnetization*, and B is defined on the whole space $\mathbb{R}^3$ (remember $|y| \leq \frac{|x|}{2}$ implies $|x-y| \geq ||x| - |y|| \geq \frac{1}{2}|x|$), i.e., in the case of a continuous magnetization m the potential V in (3.114) is regular at infinity:

$$
B(x) = O\left(\frac{1}{|x|^2} \right), \quad |x| \to \infty. \tag{3.115}
$$

The *dipole magnetic induction* is the gradient field of a dipole potential B of a distribution of magnetization m

$$
\begin{aligned}
b(x) &= \nabla_x \frac{\mu_0}{4\pi} \int_B m(y) \cdot \nabla_y \frac{1}{|x-y|} \, dy & (3.116) \\
&= -\nabla_x \frac{\mu_0}{4\pi} \int_B \frac{m(y) \cdot (x-y)}{|x-y|^3} \, dy, \quad x \in \mathbb{R}^3.
\end{aligned}
$$

For points $x \in \mathbb{R}^3 \backslash \overline{\mathcal{B}}$ we have

$$b(x) = -\frac{\mu_0}{4\pi} \int_{\mathcal{B}} \nabla_x \frac{m(y) \cdot (x-y)}{|x-y|^3} \, dy \tag{3.117}$$

$$= -\frac{\mu_0}{4\pi} \int_{\mathcal{B}} \left(\frac{m(y)}{|x-y|^3} - 3 \frac{(x-y) \cdot m(y)}{|x-y|^5} (x-y) \right) dy.$$

The magnetic field b is divergence-free outside $\mathcal{B}$, i.e., for points $x \in \mathbb{R}^3 \backslash \overline{\mathcal{B}}$

$$\nabla \cdot b(x) = \nabla \cdot \nabla B(x) = \Delta B(x) = 0, \quad x \in \mathbb{R}^3 \backslash \overline{\mathcal{B}}. \tag{3.118}$$

For points $x \in \mathbb{R}^3 \backslash \overline{\mathcal{B}}$, we are able to deduce that

$$\int_{\mathcal{B}} m(y) \cdot \nabla_y \frac{1}{|x-y|} \, dy = \int_{\mathcal{B}} \left(\nabla_y \cdot \frac{m(y)}{|x-y|} - \frac{1}{|x-y|} \nabla_y \cdot m(y) \right) dy. \tag{3.119}$$

In connection with (3.114) and the theorem of Gauss, we obtain the following result:

Let m be continuously differentiable on $\overline{\mathcal{B}}$. Then, for $x \in \mathbb{R}^3 \backslash \overline{\mathcal{B}}$, the potential

$$B(x) = \frac{\mu_0}{4\pi} \int_{\mathcal{B}} m(y) \cdot \nabla_y \frac{1}{|x-y|} \, dy \tag{3.120}$$

can be expressed by a volume and surface integral of the form

$$B(x) = -\frac{\mu_0}{4\pi} \int_{\mathcal{B}} \frac{1}{|x-y|} \nabla_y \cdot m(y) \, dy \tag{3.121}$$

$$+ \frac{\mu_0}{4\pi} \int_{\partial \mathcal{B}} \frac{1}{|x-y|} m(y) \cdot \nu(y) \, dS(y),$$

where ν is the (unit) normal field directed outward to $\overline{\mathcal{B}}$.

We are led to the already known consequence that the dipole magnetic induction $b = \nabla B$ in free space outside a magnetized body $\overline{\mathcal{B}}$ provides information about the divergence of the magnetization, rather than the magnetization itself. Thus, any magnetization distribution that is divergence-free is magnetically silent, so that it cannot affect the external magnetic field. Moreover, the decomposition (3.121) into a sum of surface integral and volume integral shows that the contribution to the potential B from $\overline{\mathcal{B}}$ only comes from sources and sinks of the magnetization in $\mathcal{B}$, where $\nabla \cdot m \neq 0$, and from parts of the surface $\partial \mathcal{B}$, where the surface normal ν and m are not orthogonal.

3.5 Gravito-Magneto Combined Potential Relation

Usually, in geoscientific applications, gravity and magnetic field inversion are operated separately (see e.g., [10, 18, 138]) to achieve density and magnetization independently of each other.

"Poisson's relation" (cf. [18]) shows, however, that for each regular subset $\mathcal{K}$ of a regular region $\mathcal{B} \subset \mathbb{R}^3$ with uniform magnetization (i.e., $m(y) = m$, $y \in \overline{\mathcal{K}}$) and uniform density (i.e., $\rho(y) = \rho$, $y \in \overline{\mathcal{K}}$)

$$B(x) = \frac{\mu_0}{4\pi} m \cdot \int_{\mathcal{K}} \nabla_y \frac{1}{|x - y|} \, dy = -\frac{\mu_0}{4\pi} m \cdot \int_{\mathcal{K}} \nabla_x \frac{1}{|x - y|} \, dy \tag{3.122}$$

$$= -\frac{\mu_0}{4\pi} \frac{m}{\rho} \cdot \int_{\mathcal{K}} \rho(y) \nabla_x \frac{1}{|x - y|} \, dy = -\frac{\mu_0}{\rho} \frac{m}{\rho} \cdot \nabla_x V(x) = -\frac{\mu_0}{\rho} m \cdot v(x)$$

applies to all $x \in \mathbb{R}^3 \setminus \overline{\mathcal{K}}$, where V is the Newtonian volume potential.

In other words, we are led to the following conclusion of importance for a gravito-magneto combined inversion: For a body of uniform density as well as uniform magnetization, the magnetic potential is proportional to the gravitational field component in the direction of magnetization.

3.6 Susceptibility and Permeability

Next we follow the standard literature in geomagnetics, in order to characterize the macroscopic framework. The principle to be realized is that, at distances large in comparison to the spatial extension, every current distribution generates a magnetic induction, that can be approximately understood as a magnetic dipole:

Comparing (3.99) with (3.82), we are able to replace the current density j by an effective current density $j_m = \nabla \wedge m$. So, all current distributions may be considered as consisting of two parts, namely j_m (magnetization current) and j (free current). In other words, we are led to write the basic equation for b as follows:

$$\nabla \wedge b = \mu_0(j + j_m) = \mu_0(j + \nabla \wedge m). \tag{3.123}$$

This equation can be rewritten in the form

$$\nabla \wedge \underbrace{\left(\frac{1}{\mu_0} b - m\right)}_{=h} = j, \tag{3.124}$$

where a new macroscopic field, h, called the *magnetic field* is introduced by

$$h = \frac{1}{\mu_0}(b - m). \tag{3.125}$$

As a consequence, the macroscopic equations are given by the pre-Maxwell equations

$$\nabla \wedge b = j, \tag{3.126}$$

$$\nabla \cdot b = 0, \tag{3.127}$$

i.e.,

$$\nabla \wedge b = \mu_0 j + \nabla \wedge m, \tag{3.128}$$

$$\nabla \cdot b = 0. \tag{3.129}$$

Macroscopic magnetostatics is characterized by a relationship between b and h, which according to the formula (3.125) is equivalent to a relationship between m and h (or b). In fact, for a large number of materials, there is an approximate linear relationship between m and h.

If an object made of magnetically linear and additionally isotropic material is placed in a magnetic field produced by a distant electromagnet, the magnetization $m(x)$ at a source point x may be assumed to be the product of the *magnetic susceptibility* $\chi(x)$ and the magnetic field $h(x)$, i.e.,

$$m(x) = \chi(x)\, h(x). \tag{3.130}$$

The magnetic susceptibility is called *paramagnetic*, if $\chi > 0$; the magnetic susceptibility is called *diamagnetic*, if $\chi < 0$. Usually, χ is quite small for paramagnetic and diamagnetic materials (i.e., $|\chi| \approx 10^{-5}$).

A linear relationship between h and m also implies a linear relationship between b and h. In fact, the magnetic dipole induction field b at the same source point x is given by

$$b(x) = \mu_0\, (h(x) + m(x)), \tag{3.131}$$

where μ_0 is understood to be the *permeability of free space*. Consequently, we can express (3.130) in terms of the susceptibility χ by substituting (3.130) into (3.131) so that

$$b(x) = \mu_0\, (1 + \chi(x))\, h(x) \tag{3.132}$$
$$= \mu_0\, \mu_{\text{rel}}(x)\, h(x)$$
$$= \mu(x)\, h(x),$$

where *the relative permeability* $\mu_{rel}(x)$ is defined by

$$\mu_{\text{rel}}(x) = 1 + \chi(x) \tag{3.133}$$

and *the absolute permeability* is given by

$$\mu(x) = \mu_0 \, \mu_{\text{rel}}(x) = \mu_0 \, (1 + \chi(x)). \tag{3.134}$$

Summarizing our considerations, we are led to the following conclusions (see also [27, 29]):

Magnetization is a vector quantity defined as magnetic dipole moment per unit volume. The magnetization, m, at any location within a magnetic body B is related to the field h at that location through the magnetic susceptibility, χ, by $m = \chi \, h$. In other words, the more susceptible a material, the more magnetized it becomes when placed in an inducing field. Another magnetic quantity is the dipole magnetic field, b, which is related to the field h through the magnetic permeability, μ, by $b = \mu \, h$. Permeability is related to susceptibility by $\mu(x) = \mu_0 \, (1 + \chi(x))$, where μ_0 is the permeability in free space. In vacuum, we have $\mu = 1$ and in air, we have $\mu \simeq 1$.

There essentially are two characterizations, whereby matter may acquire a macroscopic magnetic moment distribution (once more, by macroscopic we mean averaged over a large number of atoms), namely paramagnetic for $\chi > 0$ and diamagnetic for $\chi < 0$.

In magnetic prospection, the quantity measured is b (or some functionals of it), although we are often interested in the field h. However, since usually $\mu \simeq 1$ we treat b as if it were h.

Altogether, following the conventions in dipole magnetic theory, the understanding of the three mechanisms of magnetization is as follows (cf. Fig. 3.9):

- (*Diamagnetism*) The application of a magnetic induction to a diamagnetic medium induces currents within the atomic systems, and these in turn lead to a macroscopic magnetic moment density opposite in direction to the applied field (cf. Fig. 3.9a).
- (*Paramagnetism*) Atoms of paramagnetic materials have non-zero moments without the presence of the external field and the magnetic susceptibility of such materials is positive. The direction of magnetization of individual atoms is

Fig. 3.9 Schematic diagram showing the orientation of magnetic moments in the crystal lattice of different materials: (a) diamagnetic, (b) paramagnetic, and (c) ferromagnetic, modified illustration from [139] following [101]

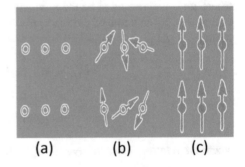

(a)   (b)   (c)

Fig. 3.10 Schematic diagram
showing the orientation of
magnetic moments in the
crystal lattice of different
materials: (c) ferromagnetic,
(d) antiferromagnetic, (e)
ferrimagnetic, modified
illustration from [139]
following [101]

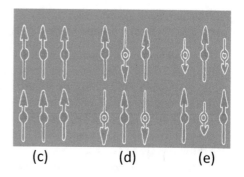

randomly oriented and their vector sum is non-zero, but weak. In presence of an
external field, the magnetic atom slightly aligns forming a weak magnetization–
an induced magnetization. When the external field is removed, the magnetization
ceases (cf. Fig. 3.9b).

- (*Ferromagnetism*) Certain paramagnetic materials have such strong magnetic
  interactions that the magnetic moments in large "domains" align. This effect is
  called ferromagnetism and is more than 100 times the effect of diamagnetism.
  So, within solid ferromagnets, there are very strong quantum-mechanical forces
  tending to make the intrinsic angular momenta of neighboring atoms line up.
  These results in domains of macroscopic size having net magnetizations (cf.
  Fig. 3.9c). The ferromagnetism decreases with increasing temperature and ceases
  when temperature exceeds the Curie point.

Some materials have domains further divided into subdomains with opposite
orientation, and the overall magnetic moment nearly cancels. These materials are
called antiferromagnetic (see Fig. 3.10d) and their susceptibility is low. There is
a last group having subdomains also aligned in oppositions; however, their net
magnetic moment is non-zero. This could be either due to the fact that one
orientation of subdomains has a weaker moment or that there are less domains with
one of orientations (see also [139]). Such substances are called *ferrimagnetic* (see
Fig. 3.10e).

Summarizing we are led to the following conclusion:

Magnetization is the vector field that expresses the density of permanent or
induced magnetic dipole moments in a magnetic material. The induced magnetization is directly proportional to the susceptibility and the concentration of magnetic
minerals present in the material. The orientation is, of course, the same as that
of the external field (in our work, the geomagnetic field). However, the measured
magnetization is not always of this direction. Responsible for this phenomena is the
remanent magnetization. The remanent magnetization is present even if we remove
the external magnetic field.

For a more detailed specification of the magnetization in rocks and minerals the
reader is referred, e.g., to [30, 144] and the literature therein. More background

information is, e.g., due to [18, 101]. The most common types of remanent magnetization can be found in a classification, e.g., following [139]:

- *Thermoremanent magnetization* is created when magnetic material is cooled below the Curie-temperature in the presence of external magnetic field (usually the Earth's magnetic field). Its direction depends on the direction of the external field at the time and place where the rock cooled.
- *Detrital magnetization* has fine-grained sediments. When magnetic particles slowly settle, they are oriented into a direction of an external field. Various clays exhibit this type of remanence.
- *Chemical remanent magnetization* is created during the growth of crystals or during an alteration of existing minerals. The temperature must be low (below the Curie point). This type might be significant in sedimentary or metamorphic rocks.
- *Isothermal remanent magnetization* is the residual left following the removal of an external field. Its amplitude is low unless it was created within very large magnetic field like during the lightning strike.
- *Viscous remanent magnetization* is produced by a long exposure to an external field. It grows with a logarithm of time. It is common for all rock types, the direction is usually close to the direction of present magnetic field, is quite stable, and an amplitude could be up to 80% of the induced magnetization.
- *Dynamic remanent magnetization* is created when a rock is exposed to variate pressures within a magnetic field. The pressures could be of various types ranging from tectonic or seismic pressures up to hammer strikes.

Chapter 4
Inverse Magnetometry

A geomathematically reflected survey about inverse problems is the subject of many publications. In the approach presented here we follow [59], thereby only considering such parts that are necessary prerequisites for our regularization strategy, i.e., multi-scale decorrelation by mollification.

4.1 Ill-Posedness and Regularization Methods

Let X and Y be Hilbert spaces and let $A : X \to Y$ be a bounded linear operator whose range $\mathcal{R}(A)$ is not necessarily closed.

Then we have the orthogonal decompositions

$$X = \mathcal{N}(A) \oplus \mathcal{N}(A)^{\perp}, \qquad (4.135)$$

$$Y = \overline{\mathcal{R}(A)} \oplus \mathcal{R}(A)^{\perp}, \qquad (4.136)$$

and

$$\mathcal{N}(A^*) = \mathcal{R}(A)^{\perp}, \qquad (4.137)$$

where $\mathcal{N}(A)^{\perp}$ is the orthogonal complement of the null space $\mathcal{N}(A)$ of A, $\overline{\mathcal{R}(A)}$ is the closure of the range $\mathcal{R}(A)$ of A, and A^* is the *adjoint operator of* A, so that $\langle Ax, y \rangle_Y = \langle x, A^*y \rangle_X$ for all $x \in X$ and $y \in Y$.

We consider an operator equation

$$Ax = y, \quad x \in X, \ y \in Y. \qquad (4.138)$$

C. Blick et al., *Inverse Magnetometry*, Lecture Notes in Geosystems Mathematics and Computing, https://doi.org/10.1007/978-3-030-79508-5_4

Four (mutually exclusive) situations arise (see, e.g., [119]) for the following discussion:

(1) $\mathcal{R}(A)$ is dense in Y, (hence, $\mathcal{N}(A^*) = \{0\}$), and $y \in \mathcal{R}(A)$;
(2) $\mathcal{R}(A)$ is dense in Y, and $y \notin \mathcal{R}(A)$;
(3) $\overline{\mathcal{R}(A)}$ is a proper subspace of Y, and $y \in \mathcal{R}(A) + \mathcal{R}(A)^\perp$;
(4) $\overline{\mathcal{R}(A)} \neq Y$, and $y \notin \mathcal{R}(A) + \mathcal{R}(A)^\perp$.

In case (1) one has, of course, a solution in the classical sense; in case (2) and (4) a classical solution does not exist, while in case (3) a solution need not exist.

An element x is called a *"least-squares solution"* of (4.138) if (see, e.g., [36, 117, 118])

$$\inf\{\|Au - y\| : u \in X\} = \|Ax - y\|. \tag{4.139}$$

Since

$$\|Au - y\|^2 = \|Au - Qy\|^2 + \|y - Qy\|^2, \tag{4.140}$$

where Q is the orthogonal projector of Y onto $\overline{R(A)}$, it is clear that a least-squares solution exists if and only if

$$y \in \mathcal{R}(A) + \mathcal{R}(A)^\perp, \tag{4.141}$$

where $\mathcal{R}(A) + \mathcal{R}(A)^\perp$ is a dense set in Y. For such y the set of all least-squares solutions of (4.138), denoted by $\mathcal{L}(y)$, is known from every textbook of numerical analysis to be a nonempty closed convex set (indeed, $\mathcal{L}(y)$ is the translate of $\mathcal{N}(A)$ by a fixed element of $\mathcal{N}(y)$), hence, it has a unique element of minimal norm, denoted by $A^\dagger y$.

In accordance to the classification due to Hadamard [80, 81] we call a problem (4.138) *well-posed*, if the following properties are valid:

- For all data, a solution exists *(existence)*.
- For all data, the solution is unique *(uniqueness)*.
- The solution depends continuously on the data *(stability)*.

In the language of functional analysis, these requirements can be translated into the following properties:

- A is surjective, i.e., the range $\mathcal{R}(A)$ is equal to Y.
- A is injective, i.e., the null space $\mathcal{N}(A)$ only consists of $\{0\}$.
- A^{-1} is continuous (i.e., bounded).

By convention, if one of the three conditions is not fulfilled, the problem (4.138) is called *ill-posed in the sense of Hadamard*.

We point out the consequences of the violations of the above requirements for the operator Eq. (4.138). The lack of injectivity of A is perhaps the easiest problem. The space X can be replaced by the orthogonal complement $\mathcal{N}(A)^\perp$, and the restriction of the operator A to $\mathcal{N}(A)^\perp$ leads to an injective problem.

In geoscientific practice, one is very often confronted with the problem that $\mathcal{R}(A) \neq Y$, since the right side is given by measurements and is, therefore, disturbed by errors. In this case, instead of $y \in \mathcal{R}(A)$, we have to consider a perturbed right side y^ε. We suppose that

$$\|y - y^\varepsilon\|_Y \leq \varepsilon. \tag{4.142}$$

The aim now is to find a solution x^ε of the equation

$$Ax^\varepsilon = y^\varepsilon. \tag{4.143}$$

Since y^ε might not be in $\mathcal{R}(A)$, the solution of this equation might not exist, and we have to generalize what is meant by a solution.

The *generalized inverse (or pseudoinverse)* $A^\dagger$ is the linear operator which assigns to each $y \in \mathcal{D}(A^\dagger) = \mathcal{R}(A) + \mathcal{R}(A)^\perp$, the unique element in the set $\mathcal{L}(y) \cap \mathcal{N}(A)^\perp$, so that $\mathcal{L}(y) = A^\dagger y + \mathcal{N}(A)$. It is easy to show that $A^\dagger y$ is the minimal norm least-squares solution, equivalently, the unique solution in $\mathcal{N}(A)^\perp$ of the *normal equation*

$$A^*Ax = A^*y. \tag{4.144}$$

It also follows that $A^\dagger = (A|_{\mathcal{N}(A)^\perp})^{-1}Q$ so that $A^\dagger$ can be characterized as the linear operator with the properties:

$$\mathcal{D}(A^\dagger) = \mathcal{R}(A) + \mathcal{R}(A)^\perp, \quad \mathcal{N}(A^\dagger) = \mathcal{R}(A)^\perp = \mathcal{N}(A^*) \tag{4.145}$$

and

$$\mathcal{R}(A^\dagger) = \mathcal{N}(A)^\perp. \tag{4.146}$$

The equivalence of these characterizations of $A^\dagger$ is established in [117, 119] (see also [120, 123–125] for a comprehensive theory of various generalized inverses and regularization methods).

In case (1) above, $A^\dagger$ gives the minimal norm solution to (3). In case (3), Eq. (4.138) has a least-squares solution (which is unique if and only if $\mathcal{N}(A) = \{0\}$). In both cases, the infimum in (4.139) is attained and is equal to zero and $\|y - Qy\|_Y$, respectively. Case (2) and (4) are pathological and usually are not under discussion in generalized inverse theory, since in both cases $y \notin \mathcal{D}(A^\dagger)$, and the infimum in (4.139) is not attained.

For gravimetric as well as magnetometric considerations and gradiometric satellite problems in geosciences (as pointed out in [37, 42, 63, 71], see also the references therein), the cases (1)–(3) are of practical interest. Case (1) corresponds to identification with complete and exact information. Case (2) may correspond to the identification in the presence of contamination in the measurements. In either of these cases, it is theoretically possible to approximate the infimum (whether actually attainable or not) to within any desired degree of accuracy. For practical reasons, it may be necessary to limit the accuracy of the approximation in order to insure that certain a priori conditions are met. The cases (3) and (4) arise for more general synthesis problems if $\mathcal{N}(A^*)$ is nontrivial.

As a canonical evolution of Hadamard's classification, M. Zuhair Nashed [119] called the operator equation (4.138) *well-posed in the least-squares (relative to X and Y)* if, for each $y \in Y$, the equation has a unique least-squares solution (of minimal norm), which depends continuously on y; otherwise the problem is ill-posed. The advantage of adopting this notion of well-posedness is that it focuses on infinite-dimensional problems (e.g., an inconsistent finite system of linear algebraic equations will not be ill-posed in the above sense, while it is ill-posed in the sense of Hadamard). It follows immediately from the open mapping theorem in functional analysis that the following statements are equivalent:

(α) The problem (4.138) is well-posed in the sense of Nashed;
(β) $\mathcal{R}(A)$ is closed;
(γ) $A^\dagger$ is bounded.

Summarizing we are led to the following conclusion (see [119, 120]):

The problem $(A; X, Y)$ is called *well-posed in the sense of Nashed*, if $\mathcal{R}(A)$ is closed in Y.

If $\mathcal{R}(A)$ is not closed in Y, the problem $(A; X, Y)$ is called *ill-posed in the sense of Nashed*.

A serious problem for ill-posed problems occurs when A^{-1} or $A^\dagger$ are not continuous. This means that small errors in the data or even small numerical noise can cause large errors in the solution. In fact, in most cases, the application of an unbounded A^{-1} or $A^\dagger$ does not make any sense in computations. The usual strategy to overcome this difficulty is to substitute the unbounded inverse operator

$$A^{-1} : \mathcal{R}(A) \longrightarrow X \tag{4.147}$$

by a suitable bounded approximation

$$R : Y \longrightarrow X. \tag{4.148}$$

The operator R usually is not chosen to be fixed, but dependent on a *regularization parameter* α.

According to the conventional approach in inverse theory, we are led to introduce the following setting:

A *regularization strategy* is a family of linear bounded operators

$$R_\alpha : Y \longrightarrow X, \quad \alpha > 0, \tag{4.149}$$

so that

$$\lim_{\alpha \to 0} R_\alpha A x = x \quad \text{for all } x \in X, \tag{4.150}$$

i.e., the operators $R_\alpha A$ converge pointwise to the identity.

From the theory of inverse problems (see, e.g., [35, 36]), it is also clear that if $A : X \to Y$ is compact and Y has an infinite dimension, then the operators R_α are not uniformly bounded, i.e., there exists a sequence $\{\alpha_j\}$ with $\lim_{j \to \infty} \alpha_j = 0$ such that

$$\|R_{\alpha_j}\|_{L(Y,X)} \to \infty, \qquad j \to \infty. \tag{4.151}$$

Here, we have used the standard convention of the operator norm

$$\|R_{\alpha_j}\|_{L(Y,X)} = \inf_{\|y\|_Y \neq 0} \frac{\|R_{\alpha_j} y\|_X}{\|y\|_Y}. \tag{4.152}$$

Note that the convergence of $R_\alpha A x$ is based on the equation $y = Ax$, i.e., on unperturbed data. In practice, the right side is affected by errors and no convergence is achieved. Instead, one is (or has to be) satisfied with an approximate solution based on a certain choice of the regularization parameter.

Let us discuss the error of the solution: We let $y \in \mathcal{R}(A)$ be the (unknown) exact right-hand side and $y^\varepsilon \in Y$ be the measured data with

$$\|y - y^\varepsilon\|_Y \leq \varepsilon. \tag{4.153}$$

For a fixed $\alpha > 0$, we let

$$x^{\alpha,\varepsilon} = R_\alpha y^\varepsilon, \tag{4.154}$$

and look at $x^{\alpha,\varepsilon}$ as an approximation of the solution x of $Ax = y$. Then the error can be split in standard way (see, e.g., [36]) as follows:

$$\|x^{\alpha,\varepsilon} - x\|_X = \|R_\alpha y^\varepsilon - x\|_X \tag{4.155}$$

$$\leq \|R_\alpha y^\varepsilon - R_\alpha y\|_X + \|R_\alpha y - x\|_X$$

$$\leq \|R_\alpha\|_{L(Y,X)} \|y^\varepsilon - y\|_Y + \|R_\alpha y - x\|_X,$$

such that

$$\|x^{\alpha,\varepsilon} - x\|_X \leq \varepsilon \|R_\alpha\|_{L(Y,X)} + \|R_\alpha Ax - x\|_X. \tag{4.156}$$

We see that the error between the exact and the approximate solution consists of two parts:

Fig. 4.11 Typical behavior
of the total error in a
regularization process

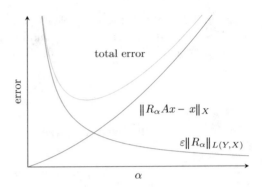

- The first term is the product of the bound for the error in the data and the norm of
  the regularization operator R_α. This term will usually tend to infinity for $\alpha \to 0$,
  if the inverse A^{-1} is unbounded (for example, if A is compact).
- The second term denotes the approximation error $\|(R_\alpha - A^{-1})y\|_X$ for the exact
  right-hand side $y = Ax$. This error tends to zero as $\alpha \to 0$ by the definition of a
  regularization strategy.

Both parts of the error show a diametrically different behavior. A typical
illustration of the errors in dependence on the regularization parameter α is sketched
in Fig. 4.11. Thus, a strategy is needed to choose α dependent on ε in order to keep
the error as small as possible, i.e., we would like to minimize

$$\varepsilon \|R_\alpha\|_{L(Y,X)} + \|R_\alpha Ax - x\|_X. \tag{4.157}$$

In principle, we distinguish two classes of parameter choice rules (see, e.g.,
[56, 59]): If $\alpha = \alpha(\varepsilon)$ does not depend on y^ε, we call $\alpha = \alpha(\varepsilon)$ an a priori
parameter choice rule. Otherwise α depends also on y^ε and we call $\alpha = \alpha(\varepsilon, y^\varepsilon)$ an
a posteriori parameter choice rule. It is conventional to say that a parameter choice
rule is convergent, if for $\varepsilon \to 0$ the rule is such that

$$\lim_{\varepsilon \to 0} \sup\{\|R_{\alpha(\varepsilon, y^\delta)} y^\varepsilon - A^\dagger y\|_X \; : \; y^\varepsilon \in Y, \; \|y^\varepsilon - y\|_Y \le \varepsilon\} = 0 \tag{4.158}$$

and

$$\lim_{\varepsilon \to 0} \sup\{\alpha(\varepsilon, y^\varepsilon) \; : \; y^\varepsilon \in Y, \; \|y - y^\varepsilon\|_Y \le \varepsilon\} = 0. \tag{4.159}$$

All in all, the rationale in most methods for resolution (approximate solvability)
of ill-posed inverse problems (see also [63]) is to construct a "solution" that is
acceptable physically as a meaningful approximation and is sufficiently stable from
the computational standpoint, hence, an emphasis must be put on the distinction
between "solution" and "resolution." The main dilemma of modeling ill-posed

problems (IPPs) is that the closer the mathematical model describes the IPP, the worse is the "condition number" of the associated computational problem (i.e., the more sensitive to errors). For ill-posed problems, the difficulty is to bring additional information about the desired solution, compromises, or new outlooks as aids to the resolution of IPP. As we saw, it is conventional to use the phrase "regularization of an ill-posed problem" to refer to various approaches to circumvent the lack of continuous dependence (as well as to bring about existence and uniqueness if necessary). Roughly speaking, this entails an analysis of an IPP via an analysis of associated well-posed problems, i.e., a system (usually a sequence or a family) of well-posed "regularizations," yielding meaningful answers to the IPP.

The *strategy of resolution and reconstruction of ill-posed problems* involves one or more of the following intuitive ideas (see [56, 59]):

- Change the notion of what is meant by a solution (e.g., ε-approximate solution: $\|A\tilde{x} - y\| \leq \varepsilon$, where $\varepsilon > 0$ is prescribed; quasi-solution: $\|A\tilde{x} - y\| \leq \|Ax - y\|$ for all $x \in \mathcal{M}$, a prescribed subset of the domain of A; least-squares solution of minimal norm, etc.).
- Modify the operator equation or the problem itself (e.g., by mollification).
- Change the spaces and/or topologies imposed on the spaces.
- Specify the type of involved noise ("strong" or "weak" noise as discussed, e.g., in [33]).

From the standpoint of mathematics (see, e.g., [56, 59] for more details), one can roughly group "regularization methods" into six categories (cf. Fig. 4.12):

- *Regularization methods in function spaces* is one category. This includes Tikhonov-type regularization, the method of quasi-reversibility, the use for certain function spaces such as scale spaces in multi-resolutions, the method of generalized inverses (pseudoinverses), e.g., in reproducing kernel Hilbert spaces.
- Resolution of ill-posed *problems by "control of dimensionality"* is another category. This includes projection methods and moment-discretization schemes. The success of these methods hinges on the possibility of obtaining an approximate solution while keeping the dimensionality of the finite dimensional problem within the "range of numerical stability." It also hinges on deriving error estimates for the approximate solutions that is crucial to the control of the dimensionality.
- A third category is constituted by *iterative methods* which can be applied either to the problem in function spaces or to a discrete version of it. The crucial ingredient in iterative methods is to stop the iteration before instability creeps into the process. Thus, iterative methods have to be modified or accelerated so as to provide a desirable accuracy by the time a stopping rule is applied.
- A fourth category is concerned with *filter methods*. Filter methods refer to procedures where, for example, values producing highly oscillatory solutions are eliminated. Various "low-pass" filters can, of course, be used. They are also crucial for the determination of a stopping rule. *Mollifier filters* are known to

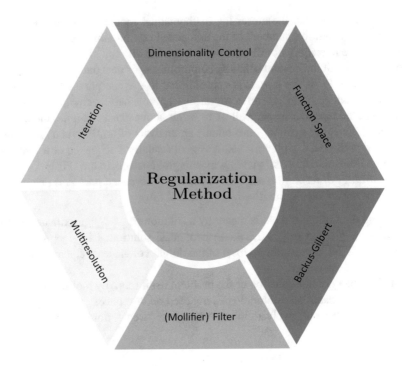

Fig. 4.12 Important regularization methods: A rough classification that does not exclude mixed/combined verifications

create sequences of smooth functions approximating a non-smooth function or a singular function. Mollification can be realized by representing a non-smooth function or a singular function in terms of frequency and/or space based function systems. The heuristic motivation is that the trouble usually comes from high-frequency components of the data and of the solution, which are damped out by filtering.

- *Multi-scale methods* use scaling function and wavelet function spaces. The analysis is usually carried out in the context of a Hilbert space for a compact operator equation. This also includes multi-scale wavelet regularization in certain Sobolev spaces. Particular examples are constituted by finite truncated singular value and infinite Tikhonov–Phillips type scale and detail spaces.
- The root of the *Backus–Gilbert method* (BG method) was geophysical (cf. [7–9]). The characterization involved in the model is known as moment problem in the mathematical literature. The BG method can be thought of as resulting from discretizing an integral equation of the first kind. Where other regularization methods, such as the frequently used Tikhonov regularization method, seek to impose smoothness constraints on the solution, the BG method instead realizes stability constraints. As a consequence, the solution is varying as little as possible if the input data were resampled multiple times. A particular feature of the BG

method is that an approximate inverse is determined independently from the right-hand side of the equation.

The *philosophy of resolution* leads to the use of algebraic methods versus function space methods, statistical versus deterministic approaches, strong versus weak noise, etc. A regularization-approximation scheme refers to a variety of methods such as Tikhonov's regularization, projection methods, multi-scale methods, iterative approximation, etc., that can be applied to ill-posed problems. These schemes turn into algorithms once a resolution strategy can be effectively implemented. Unfortunately, this requires a determination of a suitable value of a certain parameter associated with the scheme (e.g., regularization parameter, mesh size, dimension of subspace in the projection scheme, specification of the level of a scale space, classification of noise, etc.). This is not a trivial problem since it involves a trade-off between accuracy and numerical stability, a situation that does not usually arise in well-posed problems.

The original idea of a mollifier method (cf. [36, 58, 59, 95]) is interested in the solution of an operator equation, but we realize that the problem is "too ill-posed" for being able to determine the (pseudo)inverse accurately. Mollifiers (remember the comments given in the Introduction of this work) are known as smooth functions with special properties to create sequences of smooth functions approximating a non-smooth function. Thus, we provide compromises by changing the problem into a more well-posed one, namely that of trying to determine a mollified version of the solution.

The canonical mathematical cycle (cf. [44, 70]) for the treatment of inverse problems in geomathematics is as follows:

- Mathematical modeling procedure,
- mathematical analysis of the "ill-posed attitude,"
- development and execution of a mathematical regularization procedure,
- back-transfer into application and validation.

If necessary, the cycle should be applied iteratively a sufficient number of times, usually under additional input material, to achieve improvements in the solution process.

Finally, it should be mentioned that our treatment of inverse magnetometry in this book makes concrete use of earlier approaches, which include the intuitive idea of regularization through a "mollification" over a surface (cf. [64]) or a volume (cf. [47]). In terms of these concepts, the singular fundamental kernel of the "2D-Beltrami surface operator" or of the "3D-Laplace volume operator" is replaced by a suitable sequence of physically "closely related" regular kernels (see also the mathematical studies in [37, 45, 46, 49, 58–60]).

The criteria to handle the inverse magnetometry problem in our work are based on a solution philosophy that is characterized by the following three essential features:

- A (physically related) regularization property by appropriate mollification,
- A suitable wavelet decorrelation capability in detecting geological layers and structures by essentially studying surface anomalies (plus some additional information from inside),
- An economical and efficient spline and/or wavelet algorithm for modeling interesting detail interfaces in the band structures of geological strata (see [41, 42, 55, 61, 133, 134] and the references therein).

4.2   Direct and Inverse Problem Formulation

The measuring principle of magnetometry (as a counterexample to gravimetry) is shown in Fig. 4.13.

The common measuring methods provide scalar and vectorial functions of the Earth's magnetic field. By differential connection of two opposite and spatially separated magnetometers (i.e., gradiometers), tensorial gradients of the Earth's magnetic field can be measured.

Magnetometry as an inverse problem formally understood in analogy to gravimetry may be initially seen as the determination of the vectorial magnetization m from the magnetic field potential B using a scalar Fredholm integral equation

$$B(x) = \frac{\mu_0}{4\pi} \int_B \nabla_y \frac{1}{|x - y|} \cdot m(y) \, dy \qquad (4.160)$$

corresponding to the magnetic field b produced by susceptible material in $\overline{B}$. Inverse magnetometry understood in this formal sense shows similar difficulties as the gravitational case via the Newton integral equation (cf. [45, 46]).

Fig. 4.13 Principle of magnetometry (with kind permission of Teubner-Verlag taken in modified form from [86])

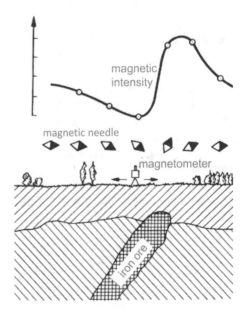

Our specific interest here is in using the mollifier methodology. It is very similar to that of decorrelative gravimetry: The magnetostatic volume integral (4.160) over $\mathcal{B}$ corresponding to a vectorial magnetization distribution m fulfills the Laplace equation in the external space $\mathcal{B}^c = \mathbb{R}^3 \backslash \overline{\mathcal{B}}$ ($\overline{\mathcal{B}} = \mathcal{B} \cup \partial\mathcal{B}$, $\partial\mathcal{B}$ boundary of $\mathcal{B}$). This property is a direct consequence of the harmonicity of the Newtonian kernel. The goal of inverse magnetometry is to determine the magnetization distribution m within a regular subarea $\mathcal{B}$ of the Earth. Usually, one assumes a vector field of limited signal energy in $\overline{\mathcal{B}}$, i.e.,

$$\|m\|_{l^2(\mathcal{B})} = \left(\int_{\mathcal{B}} |m(x)|^2 \, dx \right)^{1/2} < \infty. \tag{4.161}$$

Under this mathematical condition, the *magnetic dipole potential*

$$B(x) = \frac{\mu_0}{4\pi} \int_{\mathcal{B}} \nabla_y \frac{1}{|x - y|} \cdot m(y) \, dy, \quad x \in \mathbb{R}^3 \tag{4.162}$$

and the *dipole magnetic induction*

$$b(x) = \frac{\mu_0}{4\pi} \nabla_x \int_{\mathcal{B}} \nabla_y \frac{1}{|x - y|} \cdot m(y) \, dy, \quad x \in \mathbb{R}^3 \tag{4.163}$$

can be calculated anywhere in $\mathbb{R}^3$, so that the *direct magnetometry problem*

$$\underbrace{m}_{\text{= magnetization distribution}} \longrightarrow \underbrace{B}_{\text{= magnetic dipole potential}} \tag{4.164}$$

can be realized with the help of approximate integration (see, e.g., [51]).

As far as the point x is located in the outer space $\mathcal{B}^c = \mathbb{R}^3 \backslash \overline{\mathcal{B}}$, $B(x)$ is obtainable as regular integral. However, if the point x is located in the interior $\mathcal{B}$ or on the boundary $\partial\mathcal{B}$, we are confronted with an improper integral. Therefore, analogous to the gravimetric problem, one can assume that inverse magnetometry

$$\underbrace{B}_{\text{= magnetic dipole potential}} \longrightarrow \underbrace{m}_{\text{= magnetization distribution}} \tag{4.165}$$

also shows a specific behavior depending on the position of the input data.

In operator formulation, the integral equation (4.162) is given in the form

$$A(m) = B, \tag{4.166}$$

where the operator A can be written as volume integral over $\mathcal{B} \subset \mathbb{R}^3$

$$A(m)(x) = \mu_0 \int_{\mathcal{B}} \nabla_y G(\Delta; |x - y|) \cdot m(y) \, dy = B(x), \quad x \in \mathbb{R}^3. \tag{4.167}$$

An investigation of the ill-posedness in the sense of Hadamard is similar to that of inverse gravimetry. It can be found, e.g., in [93, 132] (see also our introductory remarks of Chap. 1).

4.3 Heuristic Perception of the Ill-Posedness

The magnetization distribution has to meet certain conditions that are quite compatible with practice. Canonically, it can be assumed that limited energy is guaranteed in the test region $\mathcal{B}$, i.e., m is mathematically a square-integrable function on $\mathcal{B}$. The function space of all vectorial square-integrable functions on the test region $\mathcal{B}$ is abbreviated with $l^2(\mathcal{B})$, so that the magnetization distribution m lies in the Hilbert function space $l^2(\mathcal{B})$.

To understand the methods for solving the inverse magnetometric problem as presented in this book, it is helpful to investigate the mathematical/physical relationship between potential and magnetization distribution in a preparatory heuristic way: We refer to the function space of all potentials B, which result as an integral over a square-integrable magnetization distribution according to

$$A(m) = B, \quad m \in X = l^2(\mathcal{B}), \ B \in Y = A(l^2(\mathcal{B})). \tag{4.168}$$

The connection between the magnetization distribution m and the potential B is mathematically expressed by the operator A, the *dipole potential operator*. This operator is assumed to "operate" between the space of magnetization distributions $X = l^2(\mathcal{B})$ and the space of potentials $Y = A(l^2(\mathcal{B}))$. In the language of functional analysis, m is the cause from the "set" of all causes X, and B is the effect from the "set" of effects Y. The operator A links cause and effect.

As we already know, in the case of a direct problem, the effect $B \in Y$ is derived from the measurable and thus known cause $m \in X$, i.e., it is calculated (directly) using the underlying operator equation (in the magnetometric case by volume integration). If the cause m cannot be measured directly or can only be measured in a very inadequate form, then one must deduce the cause m from the observed effect B. However, when using terrestrial magnetometric values, the extremely limited power of the basic equation $A(m) = B$, $m \in X$, $B \in Y$, can be explained by means of a simple, preparatory argument:

It is *not* possible to determine the magnetization distribution uniquely from measured values on the Earth's surface. In fact, it is possible to redistribute magnetization within a body in such a way that the magnetic potential does not change (see, e.g., [1, 4, 5, 18, 26, 93, 94, 107, 128, 132], and the literature therein).

Mathematically specified, we are therefore led to the following disadvantage: If we denote with m the "real magnetization distribution" of the operator equation $A(m) = B$, $m \in X$, $B \in Y$, and we are confronted with the situation that a nontrivial solution z of $Az = 0$ exists, then, because of the linearity of the operator A, we have

$$A(m + z) = A(m) + A(z) = B + 0 = B. \qquad (4.169)$$

So, z produces no effect and is, therefore, not noticeable when measuring B. In other words, z is a *"ghost,"* i.e. a phantom or artifact, when B is measured. Such ghosts frequently occur in measuring physics (for example, in inverse gravimetry), the functions z even form an infinite-dimensional space (cf. [111]). In short, the problem of the "ghosts" leads to the conclusion that in order to find the real solution m of an inverse problem one has to rely on additional information (in magnetometry, e.g., information from the inside of the test area). Moreover, the operator equation $A(m) = B$, $m \in X$, $B \in Y$, with data exclusively from the outside of $\mathcal{B}$ and/or on the surface $\partial\mathcal{B}$, is not sufficient for a reasonable solution procedure. The magnetization distribution can, therefore, only be used meaningfully if one knows additional properties about the potential B, for example, information from the inside of the subregion $\mathcal{B}$.

The appearance of ghosts, i.e., in mathematical language the uniqueness of the solution of $A(m) = B$, $m \in X$, $B \in Y$, however, is only one of the difficulties with inverse magnetometry. In addition to the occurrence of ghosts (i.e., non-uniqueness), the stability of the inversion proves to be particularly critical in practice. This is due to the fact that the data set is not available exactly, but it is affected with measurement noise. If the inverse mapping is not continuous, the solution m does not depend continuously on the data of B. As a consequence, smallest measurement errors can lead to drastic errors in a "replacement solution" or–in the worst case– make the solution completely unusable.

4.4 Magnetic Primary and Secondary Field

In the absence of an external magnetic field, individual magnetic domains within magnetic material are characterized by random directions of magnetic moments and the (net) magnetic field is zero. When placed in a primary (external) magnetic field, the magnetic moments will move toward the direction of the external field. Their orientation is no longer random and the material is said to be magnetized. The result is an induced secondary field produced by susceptible material. Its secondary magnetic field profile across the surface is shown in Fig. 4.14. Note that the *secondary field h_s* is distinct from, but caused by, the *primary inducing magnetic field h_p*.

Written in integral form, the field at a point x due to a distribution of a magnetic material within a regular region $\mathcal{B}$ is given by

$$h(x) = h_p(x) + h_s(x) \qquad (4.170)$$

$$= h_p(x) + \nabla_x \int_{\mathcal{B}} \nabla_y \, G(\Delta; |x - y|) \cdot m(y) \, dy, \quad x \in \mathbb{R}^3.$$

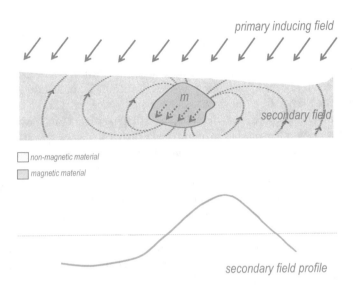

Fig. 4.14 An idealized schematic representation of the field of a magnetic body showing the primary inducing field h_p (such as the Earth's magnetic field), the induced magnetization, m, and the associated secondary field, h_s, such as the "anomalous field" produced by (ferro-)magnetic material

There are two interesting aspects in working with the secondary field, which should be mentioned:

- In practice, one usually does not work with the magnetic field, h, but with magnetic dipole field quantities such as the magnetic intensity $|b_s|$, the magnetic dipole induction b_s as a vectorial quantity, and the tensor field ∇b_s. Moreover, quantities that can be derived from the magnetic field such as oblique derivatives $l \cdot b_s$ with l an appropriately given vector field in $\mathbb{R}^3$ may be used as well.
- In accordance with the macroscopic equation (3.131)

$$b(x) = \mu_0 \left(h(x) + m(x) \right), \tag{4.171}$$

there are fields

$$b_p(x) = \mu_0 \left(h_p(x) + m(x) \right) \tag{4.172}$$

and

$$b_s(x) = \mu_0 \left(h_s(x) + m(x) \right) \tag{4.173}$$

such that

$$b_s(x) = b(x) - b_p(x) = \mu_0 \left(h(x) - h_p(x) \right) = \mu_0 \, h_s(x), \tag{4.174}$$

hence,

$$b_s(x) = \mu_0 \nabla_x \int_B \nabla_y G(\Delta; |x - y|) \cdot m(y)\, dy, \quad x \in \mathbb{R}^3. \tag{4.175}$$

In any case we are confronted with the problem that the measurements for b must be reduced by the proportion of the primary induction field b_p in a pre-phase in order to obtain the data for the secondary induction field $b_s = b - b_p$.

4.5 Born Approximation

We know that the magnetization of a body $\overline{B}$ is related to the magnetic field h at the location x as

$$m(x) = \chi(x)\, h(x) = \chi(x)\, (h_p(x) + h_s(x)). \tag{4.176}$$

If the secondary field is significant, then the magnetization is affected in both the magnitude and direction of the primary and the secondary field. If, however, the secondary field is negligible, then the body's magnetization is parallel to the inducing field with magnitude and direction proportional to the susceptibility χ. This leads over to the so-called *Born approximation* for the integral equation

$$b_s(x) = \mu_0 \nabla_x \int_B \underbrace{\nabla_y G(\Delta; |x - y|)}_{=\gamma(x,y)} \cdot \underbrace{m(y)}_{=\chi(y)h(y)}\, dy. \tag{4.177}$$

The objective now is to render a feasible subsurface distribution of the susceptibility χ, that may have given rise to the anomalous fields in the magnetic survey data. This also is an inverse problem of ill-posed nature, to which mollifier methodology is applicable in the well-known way. In geophysical practice, one often works with input data in a certain direction l, i.e., $l \cdot b_s$. If the data are acquired in free space (i.e., air), then $\mu \simeq 1$. Thus we treat $l \cdot h$ as if it were $l \cdot b$. Therefore, in connection with (3.132), we are confronted with the inverse problem of determining the susceptibility χ from scalar data of $l \cdot b_s$:

$$l(x) \cdot b_s(x) \simeq \mu_0\, (l(x) \cdot \nabla_x) \int_B (\gamma(x, y) \cdot b(y))\, \chi(y)\, dy. \tag{4.178}$$

This (Born-type) inverse problem allows a similar mollifier methodology as scalar inverse gravimetry (see [45]).

Unfortunately, when working with magnetic data collected over bodies of high susceptibility, significant demagnetization effects can occur, which make geological interpretation ambitious.

Chapter 5
Multi-Scale Inverse Mollifier Magnetometry

Mollifiers were originally presented by K.O. Friedrichs in an essay of the year 1944 (cf. [72]), which is regarded as a turning point in the modern theory of partial differential equations.

5.1 Mollifier Philosophy

In a somewhat extended understanding (cf. [45]), mollifiers today represent auxiliary functions, e.g., in constructive approximation, in order to generate sequences of smooth functions, which approximate a non-smooth (usually a singular) function. By a "mollification" the sharp features in an "irregular signature" are smoothed, but they remain close to the original non-smooth signature, which is of tremendous importance. More specifically, in our work, a mollifier is a kernel function that convolved with a particular function yields a filtered function that is "close" to the function but "smoother." In other words, in the classical approximation of a signature, a mollifier is not only one kernel function but a sequence, or usually a one-parameter (i.e., scale-dependent) family of kernels.

Although mollifiers are of particular significance in modern approximation theory, they have had no deeper importance in magnetometric exploration so far. The reason is that we need appropriately related mollifiers on two different levels, namely mollifiers on the dipole potential and the magnetization distribution level, respectively. In fact, we have to determine a contrast function, i.e., the Earth's magnetization distribution, not from data systems of the magnetization distribution itself, but from data systems of a potential, i.e., the Earth's dipole potential. This situation requires an inversion process in the framework of a Fredholm integral equation of the first kind. Thus, we need not only a mollifier sequence for the Dirac kernel, but simultaneously a mollifier sequence for the dipole potential kernel in such a way that this mollifier sequence, after inversion, has to lead back to

C. Blick et al., *Inverse Magnetometry*, Lecture Notes in Geosystems Mathematics and Computing, https://doi.org/10.1007/978-3-030-79508-5_5

the mollifier sequence for the Dirac kernel. Therefore, we need a mathematical mechanism for switching over from mollifiers of the Dirac kernel to mollifiers of the dipole potential kernel, and vice versa.

Mollifier kernels, constructed in the above described magnetometric way for two different, but appropriately related levels, turn out to become more and more space-localized, and this fact makes them extremely attractive for numerical purposes. In addition, filtering (convolution) with mollifier wavelets constituting the difference of two subsequent mollifier kernels provides space-limited band-pass information that, with increasing scales, delivers better and better details, e.g., about transitional zones between geological layers and possible irregularities.

Once more it should be mentioned that the original idea of the mollifier method in the general theory of inverse problems, already found, e.g., in the work published by [36, 95], is an "alternative solution" in which one finds that the problem is too "strongly ill-posed" to apply one of the traditional regularization methods. The heuristic motivation for the "mollification" in these early contributions is, in particular, the annoyance that high-frequency components in the solution must be specifically damped by smooth structures in order to obtain solution statements at all.

The applicability of mollifiers in this chapter is concerned with geoexploration: Our magnetometric mollifier approach is interested in associated sequences of dipole and Dirac wavelet kernels, respectively, which show decreasing space-localized supports with increasing scale for the simultaneous multi-scale approximation of both magnetometric potential and magnetization density distribution, so that numerically economic and efficient implementations become applicable. As a matter of fact, the convolutions with the inverse wavelet kernels simultaneously allow to model band-pass filtered information of the potential as well as the magnetization distribution, so that both signatures admit a certain decorrelation into detail information. The substantial idea is that, if the wavelet kernels of the magnetometric potential are constructed in close geophysical relevance, the detail information of the magnetization distribution reflects the structure of the geological layers.

The critical point, however, is the choice of the associated dipole potential and Dirac wavelet kernels. In principle, an infinite number of kernel variants can be taken for purposes of approximation. Nonetheless, we will find a reasonable compromise. It will be shown that an advantageous connection of our magneto-metric interest with appropriate tools of constructive approximation is given by the probably simplest type of a space-localizing Dirac mollifier, namely the Haar kernels. As a matter of fact, a mollifier sequence of Haar-type kernels immediately leads to a counterpart of a dipole mollifier sequence by forming taylorized versions of the dipole potential kernel, so that mollifier divergences of the vectorial dipole integral via the (negative) Laplace operator in 3D Euclidean space indeed reduce to Haar-type kernels after inversion.

As a consequence, the simultaneous approach using associated dipole potential and Dirac wavelets opens a cornucopia of perspectives (see, e.g., [45] and the references therein), for instance, signature decorrelation, noise cancelation, data compression, etc.

The disadvantage of the classical Haar kernel, however, is its discontinuity. It is constant on its support, but it vanishes outside. This often leads to numerical instabilities, especially in a marginal strip of around the boundary of a bounded exploration area. This phenomenon can be avoided, at least to some extent, by, e.g., smoothed variants of the Haar kernels.

5.2 Mollifier Framework

The inverse treatment of (4.167) can be largely performed analogous to the gravimetric case (as provided by Freeden [45]). As originally proposed by [48], we use mollifier dipole settings of radial basis function character. Radial basis functions also play a role in [2, 51, 56, 63, 79] and many other approximation procedures.

An exceptional role is played by the Haar kernels:

- The *Haar kernel of degree* 0 (simply called *Haar kernel*) is given by

$$H_\tau^0(r) = \begin{cases} \frac{1}{\|\mathbb{B}_\tau^3\|}, & r < \tau, \\ 0, & r \geq \tau, \end{cases} \tag{5.179}$$

with

$$\|\mathbb{B}_\tau^3\| = \frac{4}{3}\pi\tau^3 \tag{5.180}$$

the volume of the ball $\mathbb{B}_\tau^3$ and $G_{H_\tau^0}(\Delta; r)$ is given by

$$G_{H_\tau^0}(\Delta; r) = \begin{cases} \frac{1}{4\pi r}, & \tau \leq r \\ \frac{1}{8\pi\tau}\left(3 - \frac{1}{\tau^2}r^2\right), & r < \tau. \end{cases} \tag{5.181}$$

Equivalently,

$$G_{H_\tau^0}(\Delta; r) = \begin{cases} \frac{1}{4\pi r}, & \tau \leq r, \\ \frac{1}{4\pi r} - \frac{(\tau-r)^2 (r+2\tau)}{8\pi\tau^3 r}, & 0 < r < \tau, \\ \frac{3}{8\pi\tau}, & r = 0. \end{cases} \tag{5.182}$$

As already pointed out, the critical point of the Haar kernel of degree 0 is the discontinuity, i.e., the ordinary Haar function H_τ^0 satisfies the equation

$$- \Delta_x G_{H^0_\tau}(\Delta; |x-y|) = H^0_\tau(|x-y|) = \begin{cases} 0, & |x-y| \geq \tau, \\ \frac{3}{4\pi\tau^3}, & |x-y| < \tau. \end{cases} \quad (5.183)$$

In what follows we are therefore interested in smoothed Haar kernel variants: For $r > 0$, an ordinary 1D-Taylor expansion of the degree $n+1$ yields

$$G(\Delta; r) = \frac{1}{4\pi r} \quad (5.184)$$

$$= \frac{1}{4\pi} \sum_{k=0}^{n+1} \frac{(-1)^k}{r^{k+1}}(r-\tau)^k + O((r-\tau)^{n+2}).$$

The sum on the right side of (5.184) allows the following reformulation:

$$G(\Delta; r) = \frac{1}{4\pi} \sum_{k=0}^{n+1} \frac{(-1)^k}{r^{k+1}}(r-\tau)^k + O((r-\tau)^{n+2}) \quad (5.185)$$

$$= \frac{1}{4\pi r} - \frac{(\tau-r)^{n+2}}{4\pi r \tau^{n+2}} + O((r-\tau)^{n+2}).$$

This leads to a mollification of $G(\Delta; r) = \frac{1}{4\pi}\frac{1}{r}, r > 0$, in the form

$$G_{H^n_\tau}(\Delta; r) = \begin{cases} \frac{1}{4\pi r}, & \tau \leq r, \\ \frac{1}{4\pi r} - \frac{(\tau-r)^{n+2}((n+1)r+2\tau)}{8\pi\tau^{n+3}r}, & 0 < r < \tau, \\ \frac{n+3}{8\pi\tau}, & r = 0. \end{cases} \quad (5.186)$$

An elementary calculation shows that the kernel

$$r \mapsto G_{H^n_\tau}(\Delta; r), \quad r \in [0, \tau), \quad (5.187)$$

is a polynomial of degree $n+2$. More explicitly (cf. [19]), we have

$$G_{H^n_\tau}(\Delta; r) \quad (5.188)$$

$$= \frac{1}{8\pi\tau^{n+3}} \sum_{k=0}^{n+1} \left(\binom{n+2}{k}(n+1)(-1)^{k+1} + 2(-1)^k \binom{n+2}{k+1} \right) \tau^{n+2-k}r^k$$

$$- \frac{(n+1)(-r)^{n+2}}{8\pi\tau^{n+3}}.$$

Especially, for $n = 0$, we arrive at the kernels known from (5.179) and (5.182).

Moreover, for $\tau \leq r$, it follows that

$$-\frac{1}{r}\frac{\partial^2}{\partial r^2}(r\,G_{H^n_\tau}(\Delta;r)) = -\frac{1}{r}\frac{\partial^2}{\partial r^2}\left(r\frac{1}{4\pi r}\right) = 0, \tag{5.189}$$

whereas, for $\tau \geq r > 0$ we have

$$-\frac{1}{r}\frac{\partial^2}{\partial r^2}(r\,G_{H^n_\tau}(\Delta;r)) = -\frac{1}{r}\frac{\partial^2}{\partial r^2}\left(r\frac{1}{4\pi r} - \frac{((n+1)r+2\tau)(r-\tau)^{n+2}}{8\pi r \tau^{n+3}}\right)$$

$$= \frac{1}{r}\frac{\partial}{\partial r}\left(\frac{(\tau-r)^{n+1}(n+3)((n+1)r+\tau)}{8\pi\tau^{n+3}}\right) \tag{5.190}$$

$$= \frac{(n+1)(n+2)(n+3)}{8\pi}\frac{(\tau-r)^n}{\tau^{n+3}}.$$

In other words, we end up with the following kernel variants:

- The *Haar kernel of degree n* is given by

$$H^n_\tau(r) = \begin{cases} \frac{(n+1)(n+2)(n+3)}{8\pi}\frac{(\tau-r)^n}{\tau^{n+3}}, & r < \tau, \\ 0, & r \geq \tau. \end{cases} \tag{5.191}$$

- The *negative Laplace antiderivative of the Haar kernel of degree n* is given (modulo a harmonic function) by

$$G_{H^n_\tau}(\Delta;r) \tag{5.192}$$

$$= \begin{cases} \frac{1}{4\pi r}, & \tau \leq r, \\ \frac{1}{4\pi r} - \frac{(\tau-r)^{n+2}((n+1)r+2\tau)}{8\pi\tau^{n+3}r}, & 0 < r < \tau, \\ \frac{n+3}{8\pi\tau}, & r = 0 \end{cases}$$

or equivalently

$$G_{H^n_\tau}(\Delta;r) \tag{5.193}$$

$$= \begin{cases} \frac{1}{4\pi r}, & \tau \leq r, \\ \frac{1}{8\pi\tau^{n+3}}\sum_{k=0}^{n+1}\left(\binom{n+2}{k}(n+1)(-1)^{k+1} + 2(-1)^k\binom{n+2}{k+1}\right)\tau^{n+2-k}r^k \\ \quad -\frac{(n+1)(-r)^{n+2}}{8\pi\tau^{n+3}}, & 0 \leq r < \tau. \end{cases}$$

Of course, Haar scaling functions and wavelets in the classical form extended over one-dimensional intervals are nothing new. Basically, the essential 1D ideas can be found already in the fundamental contribution [79]. The concepts of the variants used for purposes of gravimetric exploration go back to [47]; they were worked out by Blick [19], Blick et al. [21–23], and Freeden and Nashed [57, 58, 60]. For magnetometric application, decisive concepts and ideas have been already presented in a survey chapter of the monograph [45].

Finally, the following properties, which are central in our considerations, should be mentioned:

- The smoothed Haar kernels H_τ^n satisfy the desired integral relation

$$\int_{\mathbb{R}^3} (-\Delta_y (G_{H_\tau^n}(\Delta; |x - y|))\, dy = \int_{\mathbb{R}^3} H_\tau^n(|x - y|)\, dy = 1. \qquad (5.194)$$

- Let $\mathcal{B}$ be a regular region. Assume that F is continuous in $\overline{\mathcal{B}}$. Then the limit relation

$$\lim_{\substack{\tau \to 0 \\ \tau > 0}} \int_{\mathcal{B}} H_\tau^n(|x - y|)\, F(y)\, dy = \alpha(x)\, F(x) \qquad (5.195)$$

holds true for $x \in \mathbb{R}^3$, where $\alpha(x)$ is the solid angle with respect to the surface $\partial\mathcal{B}$ at $x \in \mathbb{R}^3$.
- The function $G_{H_\tau^n}(\Delta; \cdot)$ is $(n + 1)$-times continuously differentiable everywhere.
- Let $\mathcal{B}$ be a regular region. Assume that F is continuous in $\overline{\mathcal{B}}$. Then it is easy to see (cf. [65]) that the limit relations

$$\sup_{x \in \mathcal{B}} \left| \int_{\mathcal{B}} \left(G(\Delta; |x - y|) - G_{H_\tau^n}(\Delta; |x - y|) \right) F(y)\, dy \right| = O(\tau^2) \qquad (5.196)$$

and

$$\sup_{x \in \mathcal{B}} \left| \int_{\mathcal{B}} \left(\nabla_x G(\Delta; |x - y|) - \nabla_x G_{H_\tau^n}(\Delta; |x - y|) \right) F(y)\, dy \right| \qquad (5.197)$$

$$= \sup_{x \in \mathcal{B}} \left| \int_{\mathcal{B}} \left(-\nabla_y G(\Delta; |x - y|) + \nabla_y G_{H_\tau^n}(\Delta; |x - y|) \right) F(y)\, dy \right|$$

$$= O(\tau)$$

hold true as τ tends to zero (note that we omit analogous results in L^2-topology).

The replacement of the singular Newton kernel $G(\Delta; \cdot)$ by the regular (mollifier) Haar sequence $\{G_{H_\tau^0}(\Delta; \cdot)\}_{\tau > 0}$ enables a multi-scale approximation in the form

$$B_{H_\tau^0}(x) = \mu_0 \int_B \underbrace{\nabla_y G_{H_\tau^0}(\Delta; |x-y|)}_{=\gamma_{H_\tau^0}(x,y)} \cdot m(y)\, dy, \quad x \in \mathbb{R}^3, \tag{5.198}$$

where the mollifier vectorial dipole kernel $\gamma_{H_\tau^0}(x,y)$, $x, y \in \mathbb{R}^3$, possesses the representation

$$\gamma_{H_\tau^0}(x,y) = \nabla_y G_{H_\tau^0}(\Delta, |x-y|) \tag{5.199}$$

$$= \begin{cases} \dfrac{x-y}{4\pi|x-y|^3}, & \tau_j \le |x-y|, \\[2ex] -(x-y)\left(\dfrac{(\tau-|x-y|)^2(2\tau+|x-y|)}{8\pi|x-y|^3\tau^3} - \dfrac{1}{4\pi|x-y|^3} \right. \\[2ex] \quad - \dfrac{(\tau-|x-y|)^2}{8\pi|x-y|^2\tau^3} \\[2ex] \quad \left. + \dfrac{(\tau-|x-y|)(2\tau+|x-y|)}{4\pi|x-y|^2\tau^3} \right), & 0 < |x-y| < \tau, \\[2ex] 0, & 0 = |x-y| \end{cases}$$

(for a profile illustration of $\gamma_{H_{\tau_j}^0}(x,y) = \nabla_y G_{H_{\tau_j}^0}(\Delta; |x-y|)$ see Fig. 5.15).

Moreover, as an immediate consequence, we obtain a mollification of the dipole kernel $\nabla_y G(\Delta, |x-y|)$ via the scale-dependent kernel ("vectorial scaling function") of the degree n

$$\gamma_{H_\tau^n}(x,y) = \nabla_y G_{H_\tau^n}(\Delta, |x-y|), \quad x, y \in \mathbb{R}^3, \tag{5.200}$$

in the form

Fig. 5.15 Profile lines of the first component of the dipole vector field $\gamma_{H_\tau^0}(x,y) = \nabla_y G_{H_\tau^0}(\Delta; |x-y|)$ (black) and its mollifiers in $y = 0$ for the special parameter values $\tau = 1, 0.5$, and 0.25, respectively

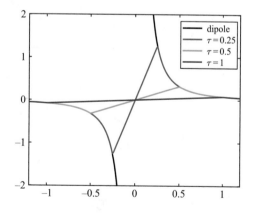

$$\gamma_{H^n_\tau}(x, y) \tag{5.201}$$

$$= \begin{cases} \frac{x-y}{4\pi|x-y|^3}, & \tau \le |x-y|, \\[2ex] -(x-y)\left(\frac{(\tau-|x-y|)^{n+2}(2\tau+|x-y|(n+1))}{8\pi|x-y|^3\tau^{n+3}} - \frac{1}{4\pi|x-y|^3}\right. \\[2ex] \quad - \frac{(\tau-|x-y|)^{n+2}(n+1)}{8\pi|x-y|^2\tau^{n+3}} \\[2ex] \quad \left. + \frac{(\tau-|x-y|)^{n+1}(2\tau+|x-y|(n+1))(n+2)}{8\pi|x-y|^2\tau^{n+3}}\right), & 0 < |x-y| < \tau, \\[2ex] 0, & 0 = |x-y|. \end{cases}$$

The profile illustrations (see Fig.5.16) show that, with increasing degree n, the extremal values of corresponding first component functions of the vector-valued kernel $\gamma_{H^n_\tau}(x, y) = \nabla_y G_{H^n_\tau}(\Delta; |x-y|)$ increase, too.

The tensorial mollifier scaling function

$$\boldsymbol{\xi}_{H^n_\tau}(x, y) = \nabla_x \gamma_{H^n_\tau}(x, y), \quad \xi, y \in \mathbb{R}^3, \tag{5.202}$$

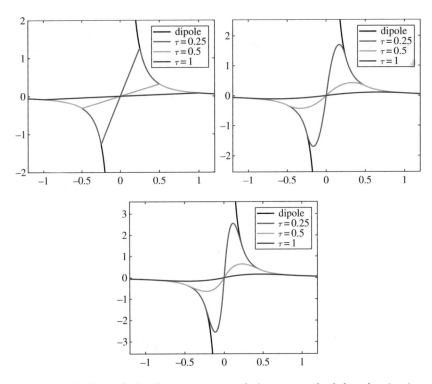

Fig. 5.16 Profile lines of the first component of the vector-valued kernel $\gamma(x, y) = \nabla_y G(\Delta; |x-y|)$ (black) and its mollifiers in $\gamma_{H^n_\tau}(x, y) = \nabla_y G_{H^n_\tau}(\Delta; |x-y|)$ in $y = 0$ for $n = 0$ (top left), $n = 1$ (top right), and $n = 2$ (bottom) and $\tau = 0.25, 0.5, 1$, respectively

is given by

$$\xi_{H_\tau^n}(x, y) = \nabla_x \gamma_{H_\tau^n}(x, y) \tag{5.203}$$

$$= \begin{cases} (x-y)(x-y)^T \dfrac{3}{4\pi|x-y|^5} - \mathbf{i}\,\dfrac{1}{4\pi|x-y|^3}, & \tau \le |x-y|, \\[2mm] -(x-y)(x-y)^T \left(\dfrac{6\tau^3(\tau-|x-y|)^n - 6\tau^{n+3}}{8\pi|x-y|^5 \tau^{n+3}} \right. \\[2mm] \quad + \dfrac{3|x-y|^3 n^2(\tau-|x-y|)^n}{8|x-y|^5 \tau^{n+3}\pi} \\[2mm] \quad + \dfrac{|x-y|^3 n^3(\tau-|x-y|)^n + 2|x-y|^3 n(\tau-|x-y|)^n}{8\pi|x-y|^5 \tau^{n+3}} \\[2mm] \quad + \dfrac{6|x-y|n\tau^2(\tau-|x-y|)^n + 3|x-y|^2 n\tau(\tau-|x-y|)^n}{8\pi|x-y|^5 \tau^{n+3}} \\[2mm] \quad \left. + \dfrac{3|x-y|^2 n^2\tau(\tau-|x-y|)^n}{8\pi|x-y|^5 \tau^{n+3}} \right) \\[2mm] \quad + \mathbf{i} \left(\dfrac{(\tau-|x-y|)^{n+2}(|x-y|+2\tau+|x-y|n)}{8\pi|x-y|^3 \tau n+3} - \dfrac{1}{4\pi|x-y|^3} \right. \\[2mm] \quad - \dfrac{(\tau-|x-y|)^{n+2}(n+1)}{8\pi|x-y|^2 \tau^{n+3}} \\[2mm] \quad \left. + \dfrac{(\tau-|x-y|)^{n+1}(n+2)(|x-y|+2\tau+|x-y|n)}{8\pi|x-y|^2 \tau^{n+3}} \right), & 0 < |x-y| < \tau, \\[2mm] -\dfrac{n^3+6n^2+11n+6}{24\pi\tau^3}, & 0 = |x-y|, \end{cases}$$

where $\mathbf{i}$ denotes the unit matrix.

5.3 Multi-Scale Mollifier Approximation

For numerical purposes, the continuous mollifier parameter τ will be replaced by a strictly monotonically decreasing sequence τ_j, $j \in \mathbb{N}_0$, with

$$\lim_{j\to\infty} \tau_j = 0, \tag{5.204}$$

and filtering has to be performed with corresponding τ_j-low- and band-pass mollifier convolutions.

In what follows, we explain the multi-scale mollifier context in a more concrete form (see, for comparison, [45] and the references therein):

The *scale-discrete vectorial wavelet kernels* $(W\gamma)_{H_{\tau_j}^n}(x, y)$, $x, y \in \mathbb{R}^3$, given by (see Fig. 5.17)

$$(W\gamma)_{H_{\tau_j}^n}(x, y) = \nabla_y G_{H_{\tau_{j+1}}^n}(\Delta; |x-y|) - \nabla_y G_{H_{\tau_j}^n}(\Delta; |x-y|) \tag{5.205}$$

$$= \gamma_{H_{\tau_{j+1}}^n}(x, y) - \gamma_{H_{\tau_j}^n}(x, y),$$

are used for the geological interpretation by multi-scale mollifier decorrelation. They also serve as a basis for the intended inversion process to deduce the magnetization distribution from geomagnetic field data.

Fig. 5.17 Profile lines of the first component of the mollifier wavelets
$(W\gamma)_{H_{\tau_j}^1}(x, y) =$
$\nabla_y G_{H_{\tau_{j+1}}^1}(\Delta; |x - y|) -$
$\nabla_y G_{H_{\tau_j}^1}(\Delta; |x - y|)$ in $y = 0$
for $\tau_j = 2^{-j}, j = 0, 1, 2,$
respectively

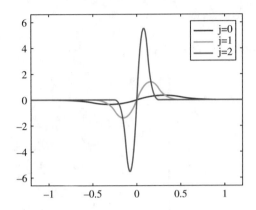

To see this in more detail we start from the dipole volume potential

$$B(x) = \mu_0 \int_{\mathcal{B}} \underbrace{\nabla_y \frac{1}{4\pi} \frac{1}{|x - y|}}_{=\gamma(x,y)} \cdot m(y)\, dy, \ x \in \mathbb{R}^3, \tag{5.206}$$

for which

$$B_{H_{\tau_j}^n}(x) = \mu_0 \int_{\mathcal{B}} \gamma_{H_{\tau_j}^n}(x, y) \cdot m(y)\, dy \tag{5.207}$$

may be understood as a *low-pass mollifier filtered dipole potential*. By taking the difference of two low-pass filtered dipole potentials for two consecutive scales $j + 1$ and j, i.e.,

$$(WB)_{H_{\tau_j}^n} = B_{H_{\tau_{j+1}}^n} - B_{H_{\tau_j}^n}, \tag{5.208}$$

we obtain the *band-pass mollifier filtered dipole potential*

$$(WB)_{H_{\tau_j}^n}(x) = \mu_0 \int_{\mathcal{B}} \underbrace{(\gamma_{H_{\tau_{j+1}}^n}(x, y) - \gamma_{H_{\tau_j}^n}(x, y))}_{=(W\gamma)_{H_{\tau_j}^n}(x,y)} \cdot m(y)\, dy. \tag{5.209}$$

For scales $J_0 \leq J - 1$, we are able to formulate the following *reconstruction formula*:

$$B_{H_{\tau_J}^n}(x) = B_{H_{\tau_{J_0}}^n}(x) + \sum_{j=J_0}^{J-1} (WB)_{H_{\tau_j}^n}(x) \tag{5.210}$$

satisfying

$$\lim_{J \to \infty} B_{H^n_{\tau_J}} = B \qquad (5.211)$$

in pointwise sense.

The J-scale low-pass mollifier filtered dipole potential $B_{H^n_{\tau_J}}$ is expressible as the sum of the J_0-scale low-pass mollifier filtered dipole potential $B_{H^n_{\tau_{J_0}}}$ and the j-band-pass mollifier filtered dipole potentials $(WB_d)_{H^n_{\tau_j}}$, $j = J_0, \ldots, J-1$. Hence, starting with the low-pass filtered dipole potential $B_{H^n_{\tau_{J_0}}}$ at a low scale J_0, we have to add more and more detail information in the form of band-pass filtered dipole potentials $(WB)_{H^n_{\tau_j}}(x)$, $j = J_0, \ldots, J-1$, until we reach the J-level low-pass mollifier filtered dipole potential $B_{H^n_{\tau_J}}$.

For sufficiently large J, we have

$$B(x) \simeq B_{H^n_{\tau_J}}(x) = \mu_0 \int_{\mathcal{B}} \gamma_{H^n_{\tau_J}}(x, y) \cdot m(y) \, dy, \quad x \in \mathbb{R}^3. \qquad (5.212)$$

In connection with a suitable cubature formula over $\overline{\mathcal{B}}$ (see, e.g., [51]) we obtain

$$B(x) \simeq B_{H^n_{\tau_J}}(x) \simeq \mu_0 \sum_{i=1}^{N_J} \gamma_{H^n_{\tau_J}}(x, y_i^{N_J}) \cdot \underbrace{w_i^{N_J} m(y_i^{N_J})}_{=a_i^{N_J}}, \quad x \in \mathbb{R}^3, \qquad (5.213)$$

where $w_i^{N_J} \in \mathbb{R}$ and $y_i^{N_J} \in \mathbb{R}^3$, $i = 1, \ldots, N_J$, are the known weights and nodes, respectively. Hence, if a finite set of potential values of the potential B is known at certain locations, the approximate formula (5.213) provides a linear system in the unknown vectorial coefficients $a_i^{N_J}$, $i = 1, \ldots, N_J$. Moreover, under the assumption that J is sufficiently large, *the decorrelation of B is obtainable by* subtracting the band-pass filtered dipole potentials

$$(WB)_{H^n_{\tau_j}}(x) \simeq \mu_0 \sum_{i=1}^{N_J} (W\gamma)_{H^n_{\tau_j}}(x, y_i^{N_J}) \cdot a_i^{N_J}, \qquad (5.214)$$

$j = J_0, \ldots, J-1$, from $B_{H^n_{\tau_J}}$.

There are several ways to obtain mollifier approximations of the magnetic induction, from which we study two approaches in more detail:

- *Tensor-Based Variant.* Since $b = \nabla B$, we are canonically led to introduce

$$b_{H^n_{\tau_j}}(x) = \nabla_x B_{H^n_{\tau_j}}(x), \quad x \in \mathbb{R}^3. \qquad (5.215)$$

This yields the *j-scale low-pass filtered magnetic dipole induction*

$$b_{H^n_{\tau_j}}(x) = \mu_0 \int_{\mathcal{B}} \xi_{H^n_{\tau_j}}(x, y)\, m(y)\, dy, \tag{5.216}$$

and the *j-scale band-pass filtered magnetic dipole induction*

$$(Wb)_{H^n_{\tau_j}} = b_{H^n_{\tau_{j+1}}} - b_{H^n_{\tau_j}}, \tag{5.217}$$

i.e.,

$$(Wb)_{H^n_{\tau_j}}(x) = \mu_0 \int_{\mathcal{B}} \underbrace{(\xi_{H^n_{\tau_{j+1}}}(x, y) - \xi_{H^n_{\tau_j}}(x, y))}_{=(W\xi)_{H^n_{\tau_j}}(x,y)}\, m(y)\, dy. \tag{5.218}$$

For integers j with $J_0 \le j \le J - 1$, this allows a decorrelation of $b_{H^n_{\tau_J}}$ via the identity

$$b_{H^n_{\tau_J}}(x) = b_{H^n_{\tau_{J_0}}}(x) + \sum_{j=J_0}^{J-1} (Wb)_{H^n_{\tau_j}}. \tag{5.219}$$

For sufficiently large J, we have

$$b(x) \simeq b_{H^n_{\tau_J}}(x) = \mu_0 \int_{\mathcal{B}} \xi_{H^n_{\tau_J}}(x, y)\, m(y)\, dy, \quad x \in \mathbb{R}^3. \tag{5.220}$$

Analogously, in connection with a suitable cubature formula over $\overline{\mathcal{B}}$, we obtain

$$b(x) \simeq b_{H^n_{\tau_J}}(x) \simeq \mu_0 \sum_{i=1}^{N_J} \xi_{H^n_{\tau_J}}(x, y_i^{N_J})\, \underbrace{w_i^{N_J}\, m(y_i^{N_J})}_{=a_i^{N_J}}, \quad x \in \mathbb{R}^3, \tag{5.221}$$

and

$$(Wb)_{H^n_{\tau_J}}(x) \simeq \mu_0 \sum_{i=1}^{N_J} (W\xi)_{H^n_{\tau_J}}(x, y_i^{N_J})\, \underbrace{w_i^{N_J}\, m(y_i^{N_J})}_{=a_i^{N_J}}, \quad x \in \mathbb{R}^3, \tag{5.222}$$

where $w_i^{N_J} \in \mathbb{R}$ and $y_i^{N_J} \in \mathbb{R}^3, i = 1, \ldots, N_J$, are the known weights and nodes, respectively. Hence, if a finite set of vector values of the field b is known at certain locations, the approximate formula (5.221) provides a linear system in the unknown coefficients $a_i^{N_J}$, $i = 1, \ldots, N_J$, and, consequently, $m(y_i^{N_J})$, $i = 1, \ldots, N_J$.

- *Scalar-Based Variant.* The field b is convolved against the scalar mollifiers $H^n_{\tau_j}$ leading to

$$b_{H^n_{\tau_j}}(x) = b_{H^n_{\tau_j}}(x) = \int_B H^n_{\tau_j}(|x-y|) \underbrace{b(y)}_{=\nabla B(y)} dy, \qquad (5.223)$$

and hence,

$$b_{H^n_{\tau_j}}(x) = \mu_0 \nabla_x \int_B \int_B H^n_{\tau_j}(|x-y|)\left(\nabla_y \frac{1}{4\pi} \frac{1}{|z-y|}\right) dy \cdot m(z)\, dz. \qquad (5.224)$$

Remark. For $x, z \in B$ the integrals

$$\iota_{H^n_{\tau_j}}(x,z) = \int_B H^n_{\tau_j}(|x-y|)\underbrace{\left(\nabla_y \frac{1}{4\pi} \frac{1}{|z-y|}\right) dy}_{=\gamma(x,y)} \qquad (5.225)$$

can be calculated in an a priori step and stored elsewhere for later use.

From (5.224) we obtain the *j-scale band-pass filtered dipole induction*

$$(WB)_{H^n_{\tau_j}}(x) = \int_B \underbrace{(H^n_{\tau_{j+1}}(|x-y|) - H^n_{\tau_j}(|x-y|))}_{=(WH)^n_{\tau_j}(|x-y|)} b(y) dy, \qquad (5.226)$$

so that, for $J_0 \le j \le J - 1$, we have

$$b_{H^n_{\tau_J}} = b_{H^n_{\tau_{J_0}}} + \sum_{j=J_0}^{J-1} (WB)_{H^n_{\tau_j}}. \qquad (5.227)$$

Application of a suitable cubature formula yields

$$b_{H^n_{\tau_J}}(x) \simeq \mu_0 \nabla_x \sum_{i=1}^{N_J} \underbrace{\int_B H^n_{\tau_J}(|x-y|)\left(\nabla_y \frac{1}{4\pi} \frac{1}{|y-y_i^{N_J}|}\right) dy}_{=\iota_{H^n_{\tau_J}}(x,y_i^{N_J})} \cdot \underbrace{w_i^{N_J} m(y_i^{N_J})}_{=a_i^{N_J}}.$$

$$(5.228)$$

Therefore, if a finite set of vector values of b is known, the cubature formula (5.228) provides a linear system in the vectorial unknowns $a_i^{N_J}, i = 1, \ldots, N_J$, and consequently the vector values $m(z_i^{N_J})$.

For $j = J_0, \ldots, J - 1$, the constituting ingredients of a multi-scale decorrelation are obtainable from

$$b_{H^n_{\tau_j}}(x) = \mu_0 \nabla_x \sum_{i=1}^{N_J} \int_{\mathcal{B}} H^n_{\tau_j}(|x - y|) \left(\nabla_y \frac{1}{4\pi} \frac{1}{|y - y_i^{N_J}|} \right) dy \cdot \underbrace{w_i^{N_J} m(y_i^{N_J})}_{=a_i^{N_J}}$$

$$\text{(5.229)}$$

and

$$(Wb)_{H^n_{\tau_j}}(x) \qquad\qquad\qquad\qquad\qquad\qquad\qquad\qquad\qquad (5.230)$$

$$= \mu_0 \nabla_x \sum_{i=1}^{N_J} \int_{\mathcal{B}} (WH)^n_{\tau_j}(|x - y|) \left(\nabla_y \frac{1}{4\pi} \frac{1}{|y - y_i^{N_J}|} \right) dy \cdot \underbrace{w_i^{N_J} m(y_i^{N_J})}_{=a_i^{N_J}}.$$

5.4   Multi-Scale Mollifier Divergence of the Magnetization

We recall (see (3.120)) that, for $x \in \mathcal{B}$,

$$b(x) = \nabla B(x) = \nabla_x \underbrace{\mu_0 \int_{\mathcal{B}} \nabla_y G(\Delta; |x - y|) \cdot m(y)\, dy}_{=B(x)} \qquad (5.231)$$

$$= \nabla_x \underbrace{\left(-\nabla_x \cdot \frac{\mu_0}{4\pi} \int_{\mathcal{B}} m(y) \frac{1}{|x - y|}\, dy \right)}_{=B(x)},$$

so that we immediately obtain from Poisson's differential equation

$$-\Delta B(x) = -\nabla \cdot \nabla B(x) = -\nabla \cdot b(x) = -\mu_0 (\nabla \cdot m(x)) \qquad (5.232)$$

and

$$-\Delta b(x) = -\nabla(\nabla \cdot b(x)) = -\mu_0 \nabla(\nabla \cdot m(x)), \qquad (5.233)$$

provided that m is Hölder continuous. Simultaneously to b we are therefore able to deduce information about the macroscopic magnetic behavior of the matter inside $\mathcal{B}$ via the discussion of the *divergence* $\nabla \cdot m$ *of the magnetization* m. The equation

$$\nabla \cdot b = \mu_0\, \nabla \cdot m \qquad\qquad\qquad\qquad (5.234)$$

should be kept in mind, since we may expect that bandlimited mollifier approximation of $\nabla \cdot b$ also provides geologic information for the test area under consideration.

For all $x \in \mathbb{R}^3$ and sufficiently large J we are able to conclude that

$$b(x) = \nabla B(x) = \nabla_x \ \underbrace{\mu_0 \int_B \nabla_y G(\Delta; |x - y|) \cdot m(y) \, dy}_{=B(x)} \qquad (5.235)$$

$$\simeq b_{H^n_{\tau_j}}(x) = \nabla B_{H^n_{\tau_j}}(x)$$

$$= \nabla_x \underbrace{\left(-\nabla_x \cdot \mu_0 \int_B m(y) \, G_{H^n_{\tau_j}}(\Delta; |x - y|) \, dy \right)}_{=B_{H^n_{\tau_j}}(x)}.$$

For $j \in \mathbb{N}_0$ we introduce the *j-scale mollifier low-pass filtered version of the divergence of the magnetization*

$$M = -\mu_0 \nabla \cdot m \qquad (5.236)$$

(cf. (5.232)) by

$$M_{H^n_{\tau_j}} = -\Delta B_{H^n_{\tau_j}}. \qquad (5.237)$$

More explicitly, we have

$$M_{H^n_{\tau_j}}(x) = -\Delta B_{H^n_{\tau_j}}(x) = -\nabla \cdot b_{H^n_{\tau_j}}(x) \qquad (5.238)$$

$$= -\nabla_x \cdot \mu_0 \underbrace{\int_B m(y) \, H^n_{\tau_j}(|x - y|) \, dy}_{=m_{H^n_{\tau_j}}(x)}$$

and

$$(WM)_{H^n_{\tau_j}}(x) = -\Delta (WB)_{H^n_{\tau_j}}(x) = -\nabla \cdot (Wb)_{H^n_{\tau_j}}(x) \qquad (5.239)$$

$$= -\nabla_x \cdot \mu_0 \underbrace{\int_B m(y) \, (WH)^n_{\tau_j}(|x - y|) \, dy}_{=(Wm)_{H^n_{\tau_j}}(x)}.$$

In other words,

$$\nabla \cdot b_{H^n_{\tau_j}}(x) = \nabla_x \cdot \mu_0 \int_B m(y) \, H^n_{\tau_j}(|x - y|) \, dy \qquad (5.240)$$

$$= \mu_0 \left(\nabla \cdot m_{H^n_{\tau_j}} \right)$$

and

$$\nabla \cdot (Wb)_{H^n_{\tau_j}}(x) = \nabla_x \cdot \mu_0 \int_{\mathcal{B}} m(y) \, (WH)^n_{\tau_j}(|x - y|) \, dy \qquad (5.241)$$

$$= \mu_0 \, (\nabla \cdot (Wm)_{H^n_{\tau_j}}),$$

where

$$(Wm)_{H^n_{\tau_j}} = m_{H^n_{\tau_{j+1}}} - m_{H^n_{\tau_j}}. \qquad (5.242)$$

In the usual manner we are able to introduce the *j-scale mollifier band-pass filtered version* $(WM)_{H^n_{\tau_j}}$ *of the divergence M of the magnetization* by letting

$$(WM)_{H^n_{\tau_j}} = M_{H^n_{\tau_{j+1}}} - M_{H^n_{\tau_j}}. \qquad (5.243)$$

Now, for sufficiently large J, we obtain from (5.235)

$$M_{H^n_{\tau_J}}(x) = M_{H^n_{\tau_{J_0}}}(x) + \sum_{j=J_0}^{J-1} (WM)_{H^n_{\tau_j}}(x), \qquad (5.244)$$

so that, in connection with (5.238), we have

$$\mu_0(\nabla \cdot m) \simeq \mu_0 \left(\nabla \cdot m_{H^n_{\tau_J}} \right) \qquad (5.245)$$

$$= \mu_0(\nabla \cdot m_{H^n_{\tau_{J_0}}}) + \mu_0 \sum_{j=J_0}^{J-1} \left(\nabla \cdot (Wm)_{H^n_{\tau_j}} \right). \qquad (5.246)$$

5.5   Mollifier Wavelet Inversion Using Pre-Information

In order to overcome the difficulties of the non-uniqueness of the solution and to avoid a full-sized coefficient matrix of the linear equation system, we can assume (as in the case of gravimetry [45]) that from terrestrial in combination with available internal preliminary information, the field b can be associated with a *known* low-pass filtered field $b_{H^n_{\tau_{J_0}}}$, $J_0 < J$.

In fact, we are led to discuss two choices to handle pre-information by a mollifier Haar field $b_{H^n_{\tau_{J_0}}}$:

- *Tensor-Based Variant.* We assume that

$$b_{H^n_{\tau_{J_0}}}(x) = \mu_0 \int_{\mathcal{B}} \boldsymbol{\xi}_{H^n_{\tau_{J_0}}}(x, y) \, m(y) \, dy \qquad (5.247)$$

is known. Then, for sufficiently large J we obtain the following expression:

$$b(x) - b_{H^n_{\tau_{J_0}}}(x) \simeq b_{H^n_{\tau_J}}(x) - b_{H^n_{\tau_{J_0}}}(x) \tag{5.248}$$

$$= \mu_0 \int_{\mathcal{B}} \left(\xi_{H^n_{\tau_J}}(x, y) - \xi_{H^n_{\tau_{J_0}}}(x, y) \right) m(y) \, dy.$$

The difference kernel occurring in the integral in (5.248) has a local support. More explicitly, for all $x \in \mathbb{R}^3$ with $|x - y| \geq \tau_{J_0}$, we have

$$\xi_{H^n_{\tau_J}}(x, y) - \xi_{H^n_{\tau_{J_0}}}(x, y) = 0. \tag{5.249}$$

In connection with a suitable cubature formula over $\overline{\mathcal{B}}$ we obtain

$$b(x) - b_{H^n_{\tau_{J_0}}}(x) \simeq b_{H^n_{\tau_J}}(x) - b_{H^n_{\tau_{J_0}}}(x) \tag{5.250}$$

$$\simeq \mu_0 \sum_{i=1}^{N_J} \left(\xi_{H^n_{\tau_J}}(x, y_i^{N_J}) - \xi_{H^n_{\tau_{J_0}}}(x, y_i^{N_J}) \right) \underbrace{w_i^{N_J} m(y_i^{N_J})}_{=a_i^{N_J}},$$

where $w_i^{N_J} \in \mathbb{R}$ and $y_i^{N_J} \in \mathbb{R}^3, i = 1, \ldots, N_J$, are the known weights and nodes, respectively.

The vectorial coefficients

$$a_i^{N_J} \in \mathbb{R}^3, \quad a_i^{N_J} = w_i^{N_J} m(y_i^{N_J}), \quad i = 1, \ldots, N_J, \tag{5.251}$$

are calculated by using the known values

$$b(x_k^{L_J}) - b_{H^n_{\tau_{J_0}}}(x_k^{L_J}), \quad x_k^{L_J} \in \mathbb{R}^3, \quad k = 1, \ldots, L_J, \tag{5.252}$$

from the linear system of equations

$$b(x_k^{L_J}) - b_{H^n_{\tau_{J_0}}}(x_k^{L_J}) \tag{5.253}$$

$$= \mu_0 \sum_{i=1}^{N_J} \left(\xi_{H^n_{\tau_J}}(x_k^{L_J}, y_i^{N_J}) - \xi_{H^n_{\tau_{J_0}}}(x_k^{L_J}, y_i^{N_J}) \right) a_i^{N_J}, \quad k = 1, \ldots, L_J.$$

The solvability of the linear system (5.253) depends essentially on the distribution and density of the data points $x_k^{L_J}, k = 1, \ldots, L_J$, for the dipole potential $B(x_k^{L_J}), k = 1, \ldots, L_J$, in relation to the prescribed knots $y_k^{N_J}$, $k = 1, \ldots, N_J$, occurring in the cubature formula:

- If the dipole potential values are available for data points $x_k^{L_J}$, $k = 1, \ldots, L_J$, in "appropriately distributed" width over $\overline{B}$, the economy of the solvability is the higher the greater J_0 can be chosen with $J_0 < J$.
- For exclusively available terrestrial values (i.e., $x_k^{L_J}$, $k = 1, \ldots, L_J$, are points of the intersection of the Earth's surface with the boundary ∂B), however, the situation $J_0 < J$, J_0 large, means a strong restriction. In fact, if one is interested in exploration at greater depths, the support has to be adapted, i.e., J_0 has to be chosen appropriately small, which of course affects the numerical economy.

As a consequence, our above approach using pre-information associated with a scale $J_0 < J$ is certainly extremely helpful numerically; however, numerical effectiveness and modeling at greater depths are mutually exclusive for terrestrial data sets. For computational purposes we have to find an appropriate compromise dependent on the available data system.

The behavior of the magnetic scalar potential in relation to vectorial magnetization distribution is similar to that of the gravitational potential in relation to the density distribution (cf. [45]).

- *Scalar-Based Variant.* We consider the mollified field

$$b_{H^n_{\tau_{J_0}}}(x) = \mu_0 \nabla_x \int_B \underbrace{\nabla_y G(\Delta; |x-y|)}_{=\gamma(x,y)} \cdot m_{H^n_{\tau_{J_0}}}(y) \, dy, \qquad (5.254)$$

with

$$m_{H^n_{\tau_{J_0}}}(y) = \int_B H^n_{\tau_{J_0}}(|y-z|) \, m(z) \, dy = \int_{\substack{|y-z| \le \tau_{J_0} \\ z \in B}} H^n_{\tau_{J_0}}(|y-z|) \, m(z) \, dz. \qquad (5.255)$$

This leads to a convolution of the vectorial magnetization m with a filtered dipole kernel in the form

$$b_{H^n_{\tau_{J_0}}}(x) = \mu_0 \nabla_x \int_B \int_B H^n_{\tau_{J_0}}(|y-z|) \, \gamma(x,y) \, dy \cdot m(z) \, dz. \qquad (5.256)$$

For sufficiently large J and $J_0 < J$ we obtain the following expression:

$$b(x) - b_{H^n_{\tau_{J_0}}}(x) \simeq b_{H^n_{\tau_J}}(x) - b_{H^n_{\tau_{J_0}}}(x) \qquad (5.257)$$

$$= \mu_0 \nabla_x \int_B \int_B H^n_{\tau_J}(|y-z|) \, \gamma(x,y) \, dy \cdot m(z) \, dz$$

$$- \mu_0 \nabla_x \int_B \int_B H^n_{\tau_{J_0}}(|y-z|) \, \gamma(x,y) \, dy \cdot m(z) \, dz.$$

Remark. The vector kernels

$$\iota_{H^n_{\tau_j}}(x, z) = \int_{\mathcal{B}} H^n_{\tau_j}(|y - z|)\, \gamma(x, y)\, dy, \quad j \in \{J_0, J\}, \tag{5.258}$$

can be evaluated in an *a priori* step. Thus, the difference kernel occurring in the integral on the right side of (5.257) represents an alternative that should be kept in mind for numerical purposes.

Finally, it should be mentioned that the magnetization distribution inside $\mathcal{B}$ can be again obtained for our scalar-based approach in already known way via linear equations generated by appropriate cubature formulas.

5.6 Mollifier Reproducing Kernel Hilbert Space Structure

We designate by

$$Z_{H^n_{\tau_j}} = S_{\gamma_{H^n_{\tau_j}}}(\mathrm{l}^2(\mathcal{B})), \quad \tau_j > 0, \; n \in \mathbb{N}_0, \tag{5.259}$$

the space of all scalar-valued *mollified dipole integrals* $F = S_{\gamma_{H^n_{\tau_j}}}(f)$ given by

$$F(x) = S_{\gamma_{H^n_{\tau_j}}}(f)(x) = \int_{\mathcal{B}} \gamma_{H^n_{\tau_j}}(x, y) \cdot f(y)\, dy, \quad f \in \mathrm{l}^2(\mathcal{B}), \tag{5.260}$$

i.e.,

$$F(x) = S_{\gamma_{H^n_{\tau_j}}}(f)(x) = \langle \gamma_{H^n_{\tau_j}}(x, \cdot), f \rangle_{\mathrm{l}^2(\mathcal{B})}, \quad f \in \mathrm{l}^2(\mathcal{B}). \tag{5.261}$$

We introduce the (scalar-valued) kernel

$$K_{Z_{H^n_{\tau_j}}}(x, y) = \int_{\mathcal{B}} \gamma_{H^n_{\tau_j}}(x, z) \cdot \gamma_{H^n_{\tau_j}}(y, z)\, dz, \quad x, y \in \mathbb{R}^3. \tag{5.262}$$

In connection with (5.262) we see that, for $a_k, a_i \in \mathbb{R}$,

$$\sum_{k=1}^{N} \sum_{i=1}^{N} a_k\, a_i \int_{\mathcal{B}} \gamma_{H^n_{\tau_j}}(x_k, z) \cdot \gamma_{H^n_{\tau_j}}(x_i, z)\, dz \tag{5.263}$$

$$= \int_{\mathcal{B}} \left| \sum_{k=1}^{N} a_k \gamma_{H^n_{\tau_j}}(x_k, z) \right|^2 dz \geq 0.$$

Moreover, the matrix

$$\left(\int_{\mathcal{B}} \gamma_{H^n_{\tau_j}}(x_k, z) \cdot \gamma_{H^n_{\tau_j}}(x_i, z)\, dz\right)_{k,i=1,\dots,N} \tag{5.264}$$

is of Gram nature, so that $K_{Z_{H^n_{\tau_j}}}(\cdot, \cdot)$ is a non-negative definite kernel, provided that the system $\{\gamma_{H^n_{\tau_j}}(x_i, \cdot)\}_{i=1,\dots,N}$ is linearly independent.

Altogether, the (scalar) image

$$Z_{H^n_{\tau_j}} = \{F : F = S_{\gamma_{H^n_{\tau_j}}}(f) = \langle \gamma_{H^n_{\tau_j}}(x, \cdot), f\rangle_{l^2(\mathcal{B})},\ f \in l^2(\mathcal{B})\} \tag{5.265}$$

$\tau_j > 0$, $n \in \mathbb{N}_0$, of the set $X = l^2(\mathcal{B})$ defines a Hilbert space $(Z_{H^n_{\tau_j}}, \langle \cdot, \cdot \rangle_{Z_{H^n_{\tau_j}}})$ possessing (5.262) as the reproducing kernel, so that, for all $F \in Z_{H^n_{\tau_j}}$, we have

$$\left\langle K_{Z_{H^n_{\tau_j}}}(x, \cdot), F\right\rangle_{Z_{H^n_{\tau_j}}} = F(x), \quad x \in \mathbb{R}^3. \tag{5.266}$$

5.7 Mollifier Spline Inversion Interpolation and Smoothing

Spline Interpolation In the space $Z_{H^n_{\tau_j}} = S_{\gamma_{H^n_{\tau_j}}}(l^2(\mathcal{B}))$, $\tau_j > 0$, interpolation may be performed in a way already known from gravimetry (cf. [45]):

Spline Interpolation for Evaluation Functionals: Suppose that there are known values $B(x_i)$, $i = 1, \dots, N$, from the dipole potential $B \in Z_{H^n_{\tau_j}}$. The minimum norm interpolation problem for determining the function S_N^B via the variational condition

$$\|S_N^B\|_{Z_{H^n_{\tau_j}}} = \min_{\substack{P \in Z_{H^n_{\tau_j}}, \\ P(x_i)=B(x_i),\ i=1,\dots,N}} \|P\|_{Z_{H^n_{\tau_j}}}$$

is well-posed in the sense that its solution S_N^B exists, is unique, and depends continuously on the data $B(x_1), \dots, B(x_N)$.

The uniquely determined solution S_N^B is explicitly given by

$$S_N^B(x) = \sum_{i=1}^{N} a_i^N\, K_{Z_{H^n_{\tau_j}}}(x_i, x), \quad a_i^N \in \mathbb{R},\ x \in \mathbb{R}^3,$$

where the (scalar) coefficients $a_1^N, \dots, a_N^N$ result as a solution of the linear system

$$\sum_{i=1}^{N} a_i^N \, K_{Z_{H_{\tau_j}^n}}(x_i, x_k) = B(x_k), \quad k = 1, \ldots, N.$$

Clearly, the interpolation procedure is designed so that the interpolating spline has the minimum $Z_{H_{\tau_j}^n}$-energy norm among all interpolators. This interpretation is again reflected by one-dimensional cubic spline interpolation, where the interpolating spline indeed shows minimal "curvature-energy" (in the linearized sense).

As in the gravimetric case (see [45] for the details), the scalar coefficients $a_i^N \in \mathbb{R}$, $i = 1, \ldots, N$, for the magnetization can be obtained by inserting

$$m_N = \sum_{i=1}^{N} a_i^N \, \gamma_{H_{\tau_j}^n}(x_i, \cdot) \tag{5.267}$$

into the dipole integral

$$S_{\gamma_{H_{\tau_j}^n}}(m_N)(x) = \int_B \gamma_{H_{\tau_j}^n}(x, y) \cdot m_N(y) \, dy \tag{5.268}$$

and then going on to the interpolation conditions

$$B(x_k) = S_{\gamma_{H_{\tau_j}^n}}(m_N)(x_k) \tag{5.269}$$

$$= \int_B \gamma_{H_{\tau_j}^n}(x_k, y) \cdot m_N(y) \, dy$$

$$= \sum_{i=1}^{N} a_i^N \int_B \gamma_{H_{\tau_j}^n}(x_k, y) \cdot \gamma_{H_{\tau_j}^n}(x_i, y) \, dy, \quad k = 1, \ldots, N.$$

Smoothing In case of *erroneous data* $\tilde{B}(x_i)$, $i = 1, \ldots, N$, spline interpolation must be replaced by a *spline smoothing* technique in the space $Z_{H_{\tau_j}^n} = S_{\gamma_{H_{\tau_j}^n}}(l^2(B))$, $\tau_j > 0$. The "smoothing spline" minimizes a functional that controls the fitting of an approximating function, under statistically oriented conditions.

For a given point system $\{x_1, \ldots, x_N\}$ the problem of adapting a smooth function to a given amount of data $(x_1, \tilde{B}_d(x_1)), \ldots, (x_N, \tilde{B}_d(x_N))$ is usually realized by looking for a potential $P \in Z_{H_{\tau_j}^n}$ so that the functional

$$\sigma_{\beta,\delta}(P) = \sum_{k=1}^{N} \left(\frac{P(x_k) - \tilde{B}_d(x_k)}{\beta_k} \right)^2 + \delta \, \|P\|_{Z_{H_{\tau_j}^n}}^2 \tag{5.270}$$

is minimized in $Z_{H_{\tau_j}^n}$, where $\beta = (\beta_1, \ldots, \beta_N)^T \in \mathbb{R}^N$ are predetermined positive weights and $\delta \geq 0$ is a parameter chosen by the user.

If possible, the values β_k, $k = 1, \ldots, N$, should be adjusted to the standard deviations of the observed data values $\tilde{B}_d(x_k), k = 1, \ldots, N$.

The parameter δ provides a measure of smoothness. A small value of δ associates confidence in the measured values, while a large value of δ does the opposite. If we formally select $\delta = 0$, smoothing will result in least-squares approximation.

If we have $\tilde{B}_d(x_k) = B_d(x_k), k = 1, \ldots, N$, we are back in spline interpolation.

Spline Smoothing for Evaluation Functionals: Let δ, $\beta_1, \ldots, \beta_N$ be given positive constants. Suppose that $(x_k, \tilde{B}_d(x_k))$, $k = 1, \ldots, N$, are given data points. Then there exists a uniquely determined spline

$$S(x) = \sum_{i=1}^{N} a_i \, K_{Z_{H_{\tau_j}^n}} (x_i, x), \quad a_i \in \mathbb{R}, \; x \in \mathbb{R}^3,$$

so that the inequality

$$\sigma_{\beta,\delta}(S) \leq \sigma_{\beta,\delta}(P)$$

applies to all $P \in Z_{H_{\tau_j}^n}$ (with equality if and only if $P = S$). S is uniquely determined by the linear system of equations

$$S(x_k) + \delta \, \beta_k^2 \, a_k = \tilde{B}_d(x_k), \quad k = 1, \ldots, N.$$

The proof of these results is standard, and it can be found, for example, in [41]. Obviously, the condition of the coefficient matrix of the linear equation system improves because of more diagonal dominance with an increasing smoothing parameter δ.

Following W. Freeden, B. Witte [67] we are also able to realize a combined spline interpolation and smoothing method. The details are omitted.

Realistic Scenario Seen from a practical point of view, however, the following scenario is more realistic: For a known vector field l with non-zero values $l(x_k), k = 1, \ldots, N$, we are interested in solving the linear system in the unknowns $a_1, \ldots, a_N$,

$$l(x_k) \cdot b(x_k) = l(x_k) \cdot \nabla_w \int_{\mathcal{B}} \gamma_{H_{\tau_j}^n} (w, z) \cdot m_N(y) \, dy|_{w=x_k} \tag{5.271}$$

$$= \sum_{i=1}^{N} a_i \, l(x_k) \cdot \nabla_w \int_{\mathcal{B}} \gamma_{H_{\tau_j}^n} (w, z) \cdot \gamma_{H_{\tau_j}^n} (x_i, z) \, dz|_{w=x_k}$$

$$+ \, \delta \, \beta_k^2 \, a_k,$$

$k = 1, \ldots, N$, to obtain the desired representation of the magnetization

$$m_N = \sum_{i=1}^{N} a_i \, \gamma_{H^n_{\tau_j}}(x_i, \cdot). \tag{5.272}$$

A particular case is the choice of known vector values

$$l(x_k) = b(x_k), \quad k = 1, \ldots N, \tag{5.273}$$

resulting in the linear system

$$|b(x_k)|^2 = b(x_k) \cdot \nabla_w \int_B \gamma_{H^n_{\tau_j}}(w, z) \cdot m_N(y) \, dy|_{w=x_k} \tag{5.274}$$

$$= \sum_{i=1}^{N} a_i \, b(x_k) \cdot \nabla_w \int_B \gamma_{H^n_{\tau_j}}(w, z) \cdot \gamma_{H^n_{\tau_j}}(x_i, z) \, dz|_{w=x_k}$$

$$+ \delta \, \beta_k^2 \, a_k,$$

$k = 1, \ldots, N.$

5.8 Mollifier Finite Pointset Wavelet Inversion

An associated mollifier formulation of (4.175), for sufficiently large J, is given by

$$b_s(x) \simeq \mu_0 \nabla_x \int_B \underbrace{\nabla_y G_{H^n_{\tau_J}}(\Delta; |x - y|)}_{= \gamma_{H^n_{\tau_J}}(x, y)} \cdot m(y) \, dy, \quad x \in \mathbb{R}^3. \tag{5.275}$$

For a discrete set of known vectorial values of the secondary field b_s we obtain a linear equation system by application of a suitable cubature formula over B (as for the gravimetric case we obtain linear relations for, e.g., $\varepsilon^3 \cdot b_s$ and $l \cdot b_s$ with $l \in \mathbb{S}^2$).

Analogously to gravimetric inversion we propose the inversion by virtue of a prescribed finite mollifier dipole system:

Let $X_{N_J} = \left\{ y_1^{N_J}, \ldots, y_{N_J}^{N_J} \right\}$, $y_i^{N_J} \neq y_k^{N_J}, i \neq k$, be a pointset of $\overline{B}$, which is suitably subdivided into the sets X_{N_j} $j = J_0, \ldots, J$, with $\sharp X_{N_j} = N_j$ and $N_{j-1} < N_j$, $j = J_0 + 1, \ldots, J$, in such a way that

$$X_{N_j} = \left\{ y_1^{N_j}, \ldots, y_{N_j}^{N_j} \right\}, \quad y_i^{N_j} = y_i^{N_J}, \quad i = 1, \ldots, N_j, \; j = J_0, \ldots, J, \tag{5.276}$$

so that

$$X_{N_{j-1}} \cup X_{N_j} = X_{N_j}, \quad j = J_0 + 1, \ldots, J, \tag{5.277}$$

$$X_{N_{j-1}} \cap X_{N_j} = X_{N_{j-1}}, \quad j = J_0 + 1, \ldots, J. \tag{5.278}$$

Assume that there are the known values

$$b_s(y_i^{N_J}) = \mu_0 \, \nabla_x \int_{\mathcal{B}} m(y) \cdot \nabla_y G(\Delta; |x - y|) \, dy \big|_{x = y_i^{N_J}} \tag{5.279}$$

for $y_i^{N_J} \in \overline{\mathcal{B}}, \; i = 1, \ldots, N_J$.

The problem is to determine *finite mollifier dipole systems*

$$(b_s)_j(x) = \mu_0 \, \nabla_x \sum_{i=1}^{N_j} m_i^{N_j} \cdot \gamma_{H_{\tau_j^n}}(x, y_i^{N_j}), \quad j = J_0, \ldots, J, \tag{5.280}$$

for the unknown magnetizations

$$m_i^{N_j} = m(y_i^{N_j}), \; i = 1, \ldots, N_j, \; j = J_0, \ldots, J, \tag{5.281}$$

inside $\mathcal{B}$ from the known values $b_s(y_i^{N_J}), i = 1, \ldots, N_J$.

Note that

$$\xi_{H_{\tau_j^n}}(x, y) = \nabla_x \, \gamma_{H_{\tau_j^n}}(x, y). \tag{5.282}$$

We consider the sums (5.280) under the vectorial interpolatory conditions

$$(b_s)_j(y_k^{N_j}) = b_s(y_k^{N_j}), \; k = 1, \ldots, N_j, \; j = J_0, \ldots, J, \tag{5.283}$$

so that the unknown coefficients $m_i^{N_j} \in \mathbb{R}^3, \, i = 1, \ldots, N_j, \, j = J_0, \ldots, J$, are determined successively as follows:

- The coefficients $m_i^{N_{J_0}}, i = 1, \ldots, N_{J_0}$, are the solutions of the system

$$b_s(y_s^{N_{J_0}}) = (b_s)_{J_0}(y_s^{N_{J_0}}) = \mu_0 \, \nabla_x \sum_{i=1}^{N_{J_0}} m_i^{N_{J_0}} \cdot \gamma_{H_{\tau_{J_0}^n}}(x, y_i^{N_{J_0}}) \big|_{x = y_s}^{N_{J_0}}, \tag{5.284}$$

$s = 1, \ldots, N_{J_0}$.

- Once the coefficients $m_i^{N_{J_0}}, \, i = 1, \ldots, N_{J_0}$, are known, we determine the coefficients $m_i^{N_{J_0+1}}, i = 1, \ldots, N_{J_0+1}$, in such a way that

$$m_i^{N_{J_0+1}} = m_i^{N_{J_0}}, \quad i = 1, \ldots, N_{J_0}. \tag{5.285}$$

Then it follows that

$$(b_s)_{J_0+1}(x) - \mu_0 \, \nabla_x \underbrace{\sum_{i=1}^{N_{J_0}} \underbrace{m_i^{N_{J_0+1}}}_{=m_i^{N_{J_0}}} \cdot \gamma_{H_{\tau_{J_0+1}^n}}(x, y_i^{N_{J_0+1}})} \qquad (5.286)$$

$$= \mu_0 \, \nabla_x \sum_{i=N_{J_0}+1}^{N_{J_0+1}} m_i^{N_{J_0+1}} \cdot \gamma_{H_{\tau_{J_0+1}^n}}(x, y_i^{N_{J_0+1}}).$$

Under the interpolatory conditions

$$b_s(y_q^{N_{J_0+1}}) = (b_s)_{J_0+1}(y_q^{N_{J_0+1}}), \quad q = 1, \ldots, N_{J_0+1}, \qquad (5.287)$$

we therefore have

$$\underbrace{(b_s)_{J_0+1}(y_q^{N_{J_0+1}})}_{=b_s(y_q^{N_{J_0+1}})} - \mu_0 \, \nabla_x \sum_{i=1}^{N_{J_0}} m_i^{N_{J_0}} \cdot \gamma_{H_{\tau_{J_0+1}^n}}(x, y_i^{N_{J_0+1}})\Big|_{x=y_q^{N_{J_0+1}}} \qquad (5.288)$$

$$= \mu_0 \, \nabla_x \sum_{i=N_{J_0}+1}^{N_{J_0+1}} m_i^{N_{J_0+1}} \cdot \gamma_{H_{\tau_{J_0+1}^n}}(x, y_i^{N_{J_0+1}})\Big|_{x=y_q^{N_{J_0+1}}}$$

for the vector unknowns $m_i^{N_{J_0+1}}, i = N_{J_0}+1, \ldots, N_{J_0+1}$.

- The subsequent steps follow from

$$\underbrace{(b_s)_j(y_q^{N_j})}_{=b_s(y_q^{N_j})} - \mu_0\nabla_x \sum_{i=1}^{N_{j-1}} \underbrace{m_i^{N_j}}_{=m_i^{N_{j-1}}} \cdot \gamma_{H_{\tau_j^n}}\left(x, y_i^{N_j}\right)\Big|_{x=y_q^{N_j}} \qquad (5.289)$$

$$= \mu_0\nabla_x \sum_{i=N_{j-1}+1}^{N_j} m_i^{N_j} \cdot \gamma_{H_{\tau_j^n}}\left(x, y_i^{N_j}\right)\Big|_{x=y_q^{N_j}},$$

$q = N_{j-1} + 1, \ldots, N_j$, where the coefficients $m_i^{N_j}, i = N_{j-1}+1, \ldots, N_j$, are the unknowns.

Finally, the knowledge of all coefficients $m_i^{N_j}, i = 1, \ldots, N_j, j = J_0, \ldots, J,$ leads to the desired multi-scale mollifier approximation of the magnetization.

5.9   Mollifier Born Optimization

Next we are concerned with a "pre-information methodology" involving susceptibility instead of magnetization distribution: We suppose that the secondary field b_s
is modelable in the form

$$(W b_s)_{H_{\tau_{J_0}}^n}^{H_{\tau_J}^n} (x) = \int_{\mathcal{B}} \nabla_x (W\gamma)_{H_{\tau_{J_0}}^n}^{H_{\tau_J}^n} (x, y) \cdot b_s(y)\, \chi(y)\, dy, \quad x \in \mathbb{R}^3, \qquad (5.290)$$

where

$$(W\gamma)_{H_{\tau_{J_0}}^n}^{H_{\tau_J}^n} (x, y) = \gamma_{H_{\tau_J}^n}(x, y) - \gamma_{H_{\tau_{J_0}}^n}(x, y). \qquad (5.291)$$

We consider the "smoothing functional"

$$\sigma_{\beta,\delta}^{\mathrm{cont}}(\chi) = \sum_{i=1}^{N} \left(\frac{\left| (W b_s)_{H_{\tau_{J_0}}^n}^{H_{\tau_J}^n} (x_i) \right| - \left| \tilde{b}_s(x_i) \right|}{\beta_i} \right)^2 + \delta \int_{\mathcal{B}} (\chi(y) - \tilde{\chi}(y))^2\, dy,$$

$$(5.292)$$

where the observed intensities of the secondary field b_s

$$\left| \tilde{b}_s(x_i) \right|, \quad x_i \in \mathbb{R}^3, i = 1, \dots, N, \qquad (5.293)$$

are given and $\tilde{\chi}$ is an observed pre-information about the susceptibility χ in $\mathcal{B}$.

We discretize $\sigma_{\beta,\delta}^{\mathrm{cont}}(\chi)$ by applying an appropriate cubature formula leading to
the functional $\sigma_{\beta,\delta}^{\mathrm{discr}}(\chi_1, \dots, \chi_N)$ given by

$$\sigma_{\beta,\delta}^{\mathrm{discr}}(\chi_1, \dots, \chi_N) \qquad (5.294)$$

$$= \sum_{i=1}^{N} \frac{1}{\beta_i^2} \left(\left| \sum_{k=1}^{L} w_k \left(\nabla_x (W\gamma)_{H_{\tau_{J_0}}^n}^{H_{\tau_J}^n} (x, y_k)|_{x=x_i} \cdot b(y_k) \right) \chi_k \right| - \left| \tilde{b}_s(x_i) \right| \right)^2$$

$$+ \delta \sum_{k=1}^{L} w_k \left| \chi_k - \tilde{\chi}(y_k) \right|^2.$$

Here, $w_k, y_k, \ k = 1, \dots, L$, are the known positive weights and the prescribed
nodes of the cubature formula, respectively, and

$$\chi_k = \chi(y_k), \quad k = 1, \dots, L, \qquad (5.295)$$

are the unknown susceptibility values for the points $y_k, \ k = 1, \dots, L$, in $\mathcal{B}$.

In other words, $(\tilde{\chi}(y_1), \ldots, \tilde{\chi}(y_L))^T$ stands for the known pre-information of the unknown susceptibility vector $(\chi(y_1), \ldots, \chi(y_L))^T$, and $|\tilde{b}_s(x_i)|$, $i = 1, \ldots, N$, characterizes the known (observed) intensities of the secondary field vectors $|b_s(x_i)|$, $i = 1, \ldots, N$, in the points $x_i \in \mathbb{R}^3$.

The values β_i, $i = 1, \ldots, N$, should be adjusted to the standard deviations of the observed data values $|\tilde{b}_s(x_i)|$, $i = 1, \ldots, N$.

The positive parameter δ serves for the smoothness control, i.e., the larger the choice of δ is, the stronger the influence of the a priori information $(\tilde{\chi}(y_1), \ldots, \tilde{\chi}(y_L))^T$ on the unknown values $(\chi(y_1), \ldots, \chi(y_L))^T$ to be calculated.

In fact, it is not hard to verify that the expression $(\sigma_{\beta,\delta}^{\mathrm{discr}}(\chi_1, \ldots, \chi_N))^2$ can be identified as the square of the $(N + L)$-dimensional vector

$$
\begin{pmatrix}
\frac{1}{\beta_1} \left| \sum_{k=1}^{L} w_k \left(\nabla_x (W\gamma)_{H_{\tau J_0}^n}^{H_{\tau J}^n}(x, y_k)|_{x=x_1} \cdot b(y_k) \right) \chi_k \right| - \left| \tilde{b}_s(x_1) \right| \\
\vdots \\
\frac{1}{\beta_N} \left| \sum_{k=1}^{L} w_k \left(\nabla_x (W\gamma)_{H_{\tau J_0}^n}^{H_{\tau J}^n}(x, y_k)|_{x=x_N} \cdot b(y_k) \right) \chi_k \right| - \left| \tilde{b}_s(x_N) \right| \\
\sqrt{\delta}\sqrt{w_1} |\chi_1 - \tilde{\chi}(y_1)| \\
\vdots \\
\sqrt{\delta}\sqrt{w_L} |\chi_L - \tilde{\chi}(y_L)|
\end{pmatrix}. \qquad (5.296)
$$

The unknown coefficients are determined in such a way that the expression $(\sigma_{\beta,\delta}^{\mathrm{discr}}(\chi_1, \ldots, \chi_L)^2$ is minimized in the set of all M-tuple $(\chi_1, \ldots, \chi_L)^T$. It should be stressed that the result of the minimization procedure is dependent on the "smoothing parameter" δ.

The mollifier wavelet

$$
x \mapsto (W\gamma)_{H_{\tau J_0}^n}^{H_{\tau J}}(x, y) = \gamma_{H_{\tau j}^n}(x, y) - \gamma_{H_{\tau J_0}^n}(x, y), \quad x \in \mathcal{B}, \qquad (5.297)
$$

is known to possess a ball as support, the width of which is determined by the scale index J_0 and thus, as already explained, has a decisive influence on the desired exploration depth. In consistency with our model specification, the scale index $J > J_0$ should be chosen as large as possible.

The non-linear minimization problem

$$
(\sigma_{\beta,\delta}^{\mathrm{discr}}(\chi_1, \ldots, \chi_L))^2 \to \min \qquad (5.298)
$$

can be solved by standard solvers in numerical analysis.

Chapter 6
Test Demonstrations

Our numerical entry test for the decorrelation of a magnetization signature is based on a well-known 2D-data set (in the (x_1, x_3)-plane), which we canonically understand as 3D-reference data set as extended to the Euclidean space $\mathbb{R}^3$ (in the direction of the x_2-axis).

6.1 Marmousi-Model

The (original 2D)-Marmousi-model is a synthetic density data set that often serves as a geological reference in seismic data processing. In 1988 it was generated by the "Geophysics Group in the Institut Francais du Petrole." It is based on a profile through the "North Quenguela Trough in the Cuanza Basin in Angola" (see Fig. 6.18).

The model is available discretely for a grid of 384×122 sample points of a distance of 24 m. It measures 9192 m in length (x_1-direction) and 2904 m in depth (x_3-direction). For more information about the Marmousi-model, the reader is referred to [104, 140]. The geological interpretation of the Marmousi-model is shown in Fig. 6.19 (according to [103]). Striking are the faults in the stratification caused by an underlying plume. Despite the fact that the (original) Marmousi-model represents a velocity model, the assumption as the divergence of the magnetization distribution is geologically justified for decorrelative validation purposes. For the test of the decorrelation capability of the 3D-Marmousi-model we accept in particular the disadvantage that it is constant in the x_2-direction, in favor of the fact that we are able to realize a comprehensive interpretation of the Marmousi-model (see Fig. 6.19).

For our test implementations we choose the Marmousi-model, also because of the lower computational effort due to the small number of data points. For numerical integration of the occurring low-pass and band-pass integrals we use cubature

© The Author(s), under exclusive license to Springer Nature Switzerland AG 2021
C. Blick et al., *Inverse Magnetometry*, Lecture Notes in Geosystems Mathematics and Computing, https://doi.org/10.1007/978-3-030-79508-5_6

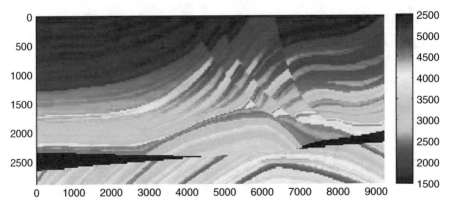

Fig. 6.18 Marmousi-model: Synthetic data set in $[\frac{\text{kg}}{\text{m}^3}]$

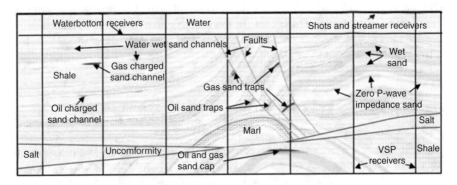

Fig. 6.19 Lithology and model characteristics including geological formations

formulas based on the concept of equidistribution in the sense of Weyl that are recruited from Euler's summation formulas (see, e.g., [51]; note that a particular multi-scale realization of equidistribution in the sense of Weyl can be found in the Ph.-D. thesis [19]). For the numerics it is advisable that the scale-parameters τ_j are adjusted to the extent of the test area. Our test area has a horizontal extent of about 10 km and is slightly less than 3 km deep with a grid point spacing of 24 m in both directions. This prompts us to select $\tau_j = 480 \cdot 2^{-j}$ m as the scale-parameter.

A schematic illustration showing the decorrelation procedure is given in Fig. 6.20. As mentioned, a magnetization m was developed whose divergence corresponds to the Marmousi data set. The identity (5.232) expresses the magnetic dipole potential by the divergence of the magnetization. This means that conclusions about geological structures in the Earth's interior can also be obtained via the divergence. The (artificial) calculations with the Marmousi-model as data template (see Fig. 6.20) confirm this assertion, and a new field of activity for the mollification method is opened in magnetic field theory.

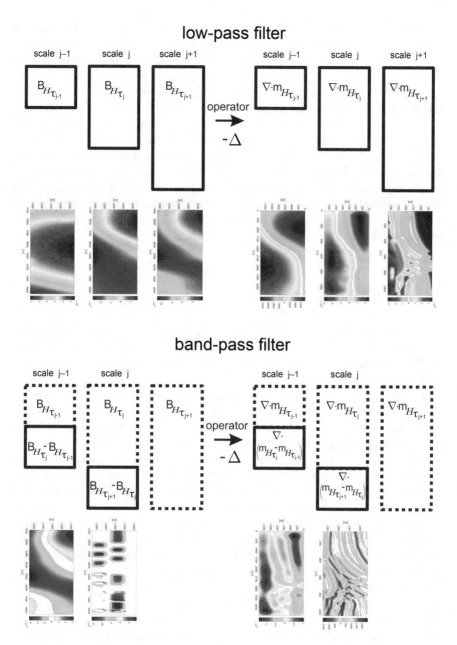

Fig. 6.20 Schematic representation of the multi-scale smoothed Haar approximation: magnetic potential (left) corresponding to the Marmousi-model as the divergence of the magnetization (right)

Looking at Fig. 6.20 in more detail, we see that its magnetic field potential does not provide useful information about the geology due to its smoothness (see Fig. 6.20, top left). However, if we proceed to the multi-scale decomposition in terms of band-pass filtered data of the dipole potential to scale j (cf. Fig. 6.20, bottom left), structural information of the magnetization of the Marmousi-model becomes visible. A key idea of the mollifier method (similar to the gravimetric case) is that the potential wavelets generating the decomposition of the potential (cf. Fig. 6.20, left) are related to the divergence of the magnetic field (cf. Fig. 6.20, top right), i.e., the magnetic field sources, via the application of the Laplace operator to Haar-type wavelets. In fact, the decorrelation of the divergence of the magnetic field in band signals (see Fig. 6.20, bottom right) shows that additional information about the separation areas of the magnetic field sources in the data set becomes available. This opens up further perspectives for the interpretation of geomagnetic field data with respect to divergence considerations.

6.2 Bavarian Molasse Area

The *Molasse basin* is a foreland basin north of the Alps that was formed during the Oligocene and Miocene epochs. The basin represents the result of the flexure of the European plate under the weight of the orogenic wedge of the Alps that was forming to the south.

The Bavarian Molasse Basin has a special significance in geothermal energy. In the central and eastern parts of the Molasse Basin, the Malm (Upper Jurassic) is the most potentially productive thermal water aquifer. Under particularly favorable circumstances, energy utilization in the Cretaceous sandstones could also be possible. Following [110], the Malm reaches the greatest thicknesses under the Molasse Basin with more than 600 m south of Munich between Lech and Inn. It is composed up to 500 m thick, massive sponge as well as light, porous coral rubble limestones deposited above them. On the northern slope, uplifted by the Bohemian Massif, several hundred meters of Malm sediments were eroded, so that today, in the eastern part of the Bavarian Molasse Basin, the thicknesses toward the Northeast decrease to less than 100 m. The movement of groundwater in the Malm is essentially along karst cavities, fractures and fault zones, and, to a lesser extent, along bed joints. The facial structure of the Malm has a different influence on the groundwater flow due to the different karstability; the facies structure of the Malm has a direct influence on its permeability. As a rule, the dolomitized mass limestones are well permeable because recrystallization led to an increase in porosity and, in addition, karstification at the grain boundaries (due to [3]).

As a consequence, most of Germany's deep geothermal systems are currently in the Malm aquifers of the Bavarian Molasse Basin. The Malm layer falls from the Danube, where it is at the surface of the Earth, toward the Alps. The geothermal relevance is caused by the fact that the Malm aquifer is filled with water due to its porosity/permeability. Locally, the Malm can be either banky or reef-like (facies) and then either fissured and/or karstified. All of these phenomena can overlap and are the subject of detailed exploration within the last decades. Because of the increasing depth toward the south, the water temperature usually increases. As a result the Malm north of Munich is only used in the heating market, south of Munich also for generating electricity. As a matter of fact, the water of the Malm generally is of drinking water quality and the salt content is very low (more details about the Bavarian Molasse Basin can be found, e.g., in [11–13, 17, 32, 112, 136, 143] and the references therein).

We test the mollifier decorrelation method on a particular data set available as seismic records for a certain area of the Bavarian Molasse Basin in the southern Germany. More explicitly, we consider the known 3D-density distribution signature of the Bavarian Molasse area as the divergence of the magnetization distribution. In fact, this (artificial) setup provides a realistic test data set varying in all the spatial directions while guaranteeing a concrete association to geological exploration. The dimension of the data set amounts to $92 \times 60 \times 58$ data points situated on a regular grid of 50 m in each direction. The vectorial test magnetization m is obtained from the scalar "data input" by discretizing the divergence operator by a finite difference scheme leading to a linear equation system to be solved for the magnetization from its corresponding divergence (note that the linear equation system is strongly under-determined, so that the decision is plausible to use artificially the additional constraint $m = 0$ on $\partial \mathcal{B}$) (Fig. 6.21).

The sectional illustration (see Fig. 6.22) shows the interaction of the occurring scaling kernel functions $\{\gamma_{H^n_{\tau_j}}(x, y)\}_{n \in \mathbb{N}_0}$ (cf. (5.207)) and wavelet kernel functions $\{(W\gamma)_{H^n_{\tau_j}}(x, y)\}_{n \in \mathbb{N}_0}$ (cf. (5.208)) in a multi-scale reconstruction of the magnetic dipole potential B of the Bavarian Molasse area. The particular scale gives information about the amount of the width of the local support of the filter functions (i.e., scaling as well as wavelet functions). The lower the scale, the larger the area covered

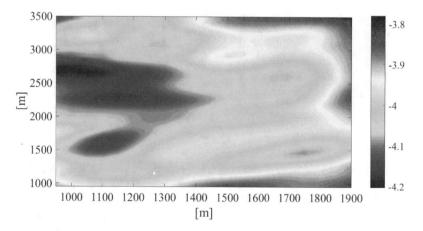

Fig. 6.21 Magnetic dipole potential B in [Tm] of the Bavarian Molasse area

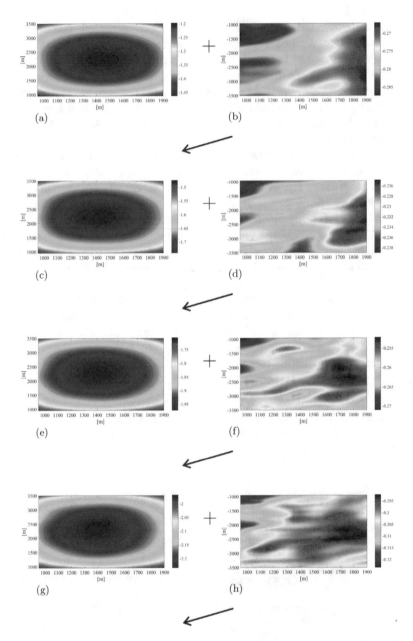

Fig. 6.22 Haar multi-scale reconstruction of the magnetic dipole potential B in [Tm]. (a) Low-pass filtering $B_{H_{\tau_0}^1}$. (b) Band-pass filtering $(WB)_{H_{\tau_0}^1}$. (c) Low-pass filtering $B_{H_{\tau_1}^1}$. (d) Band-pass filtering $(WB)_{H_{\tau_1}^1}$. (e) Low-pass filtering $B_{H_{\tau_2}^1}$. (f) Band-pass filtering $(WB)_{H_{\tau_2}^1}$. (g) Low-pass filtering $B_{H_{\tau_3}^1}$. (h) Band-pass filtering $(WB)_{H_{\tau_3}^1}$. (i) Low-pass filtering $B_{H_{\tau_4}^1}$. (j) Band-pass filtering $(WB)_{H_{\tau_4}^1}$. (k) Low-pass filtering $B_{H_{\tau_5}^1}$. (l) Band-pass filtering $(WB)_{H_{\tau_5}^1}$. (m) Low-pass filtering $B_{H_{\tau_6}^1}$. (n) Band-pass filtering $(WB)_{H_{\tau_6}^1}$. (o) Low-pass filtering $B_{H_{\tau_7}^1}$. (p) Band-pass filtering $(WB)_{H_{\tau_7}^1}$. (q) Low-pass filtering $B_{H_{\tau_8}^1}$

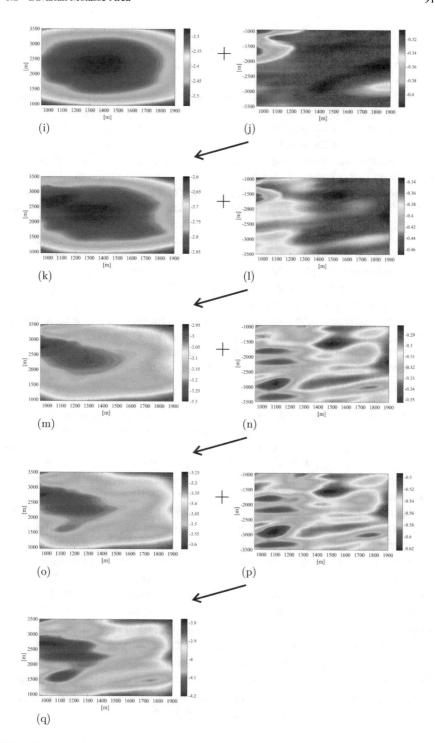

Fig. 6.22 (continued)

by the filter. As the scale increases, the width of the filter becomes smaller and smaller and the height maxima larger and larger, and thus the "filter cap" becomes steeper and steeper, so that the modeling capability of the multi-scale method changes from coarse signatures to ever finer ones during the filtering process. The aim of the multi-scale technique is thus to obtain ever more precise local information about the signature under consideration by simultaneously enlarging the data sets preferably in an equidistributed way, but not all over the test surface. Instead only local geologically relevant areas need to be under numerical investigation.

The multi-scale approximation is designed as usual by difference wavelets in such a way that a low-pass filtering to any scale j plus the band-pass filtering to the same scale j is equal to the low-pass filtering to the scale $j + 1$. Conversely, the difference between the low-pass filtering to the scale $j + 1$ and the low-pass filtering to the scale j is the band-pass filtering to the scale j. The difference of two filterings ("smoothings") results in decorrelated band signature information, which apparently provides better and better georelevant signature transitions with increasing scales.

An evaluation of the multi-scale approximation shows that only a few low-pass filterings of the dipole potential (left column in Fig. 6.22) approximate almost the entire test area sufficiently because of its "smooth" character, so that subsequent band-pass filters hardly provide any improvements. However, the multi-scale approximation (see Fig. 6.22, right column) also shows both a decorrelative space-separation into large, hardly improvable zones of the dipole potential and, in addition, geologically still significantly improvable areas, where transitional surfaces between geological layers and certain geological structures have to be prepared out.

Of eminent importance for our numerical purposes of mollifier decorrelation is that the wavelets are "space-localized" as difference functions of two successive scaling functions. In other words, they disappear outside a certain domain. These domains become simultaneously smaller as the scale increases, so that a local "zooming-in" is canonically enabled while the function maxima simultaneously increase.

Figures 6.23, 6.24, and 6.25 show a graphical representation of the described multi-scale procedure for the obtained test magnetization. Figure 6.23 illustrates the connection between the decomposition of $B_{H^n_{\tau_j}}$ and $b_{H^n_{\tau_j}}$. It turns out that the decomposition of the dipole potential $B_{H^n_{\tau_j}}$ does not yield usable information about the geology due to the smoothness of the dipole potential (Fig. 6.23, top left). However, if we go over to the multi-scale decomposition in terms of band-pass filtered data $B_{H^n_{\tau_j}} - B_{H^n_{\tau_{j-1}}}$ at scale j, structural information as detectable in the magnetization m becomes visible (Fig. 6.23, bottom left). By applying the gradient, we obtain a decomposition of $b_{H^n_{\tau_j}}$ that is still smooth and does not yield any information about the geology (Fig. 6.23, top right). However taking the difference $b_{H^n_{\tau_j}} - b_{H^n_{\tau_{j-1}}}$ (Fig. 6.23, bottom right), we are provided with vectorial geological information that, in the sum of its vector components, is comparable to the negative magnetic divergence $-\nabla \cdot m_{H^n_{\tau_{j-1}}}$ (Fig. 6.24, top right).

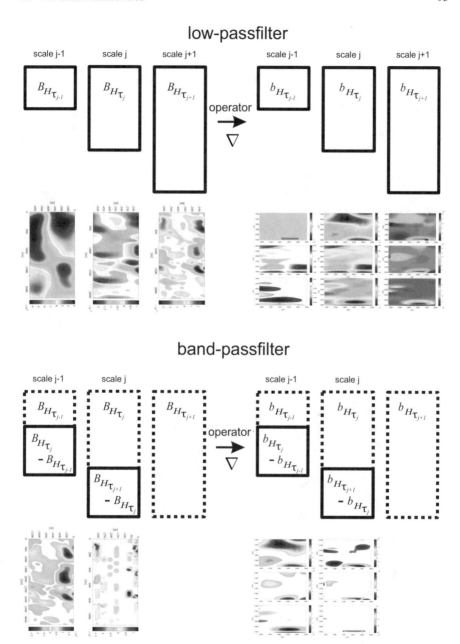

Fig. 6.23 Schematic visualization of the multi-scale mollifier decorrelation mechanisms between the geomagnetic dipole potential B in [Tm] and the field b in [T] by using the test data set from the Bavarian Molasse Basin

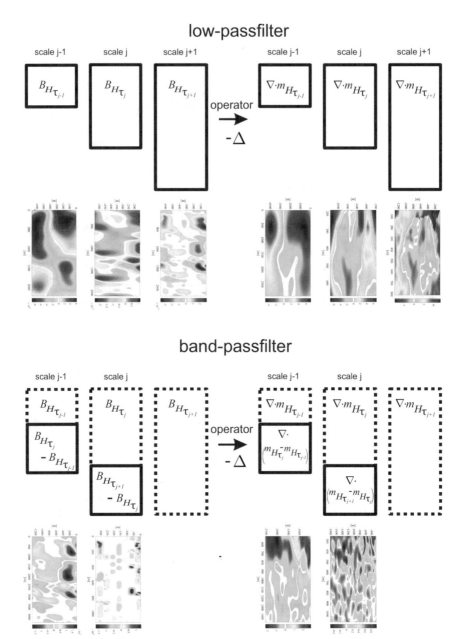

Fig. 6.24 Schematic visualization of the multi-scale mollifier decorrelation mechanisms between the geomagnetic dipole potential B in [Tm] and the divergence of the magnetization $\nabla \cdot m$ in [A/m^2] by using the test data set from the Bavarian Molasse Basin

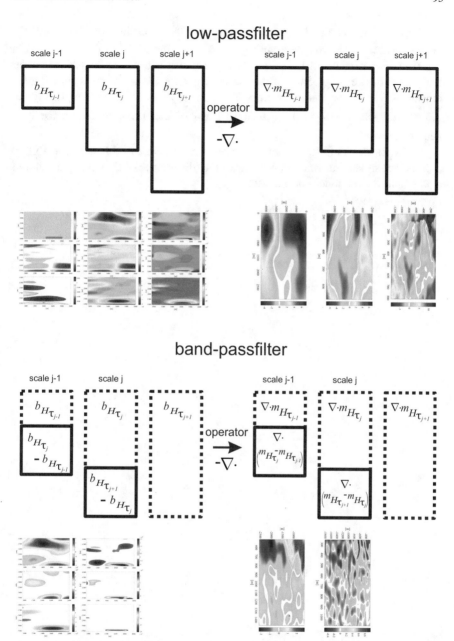

Fig. 6.25 Schematic visualization of the multi-scale mollifier decorrelation mechanisms between the geomagnetic dipole field b in [T] and the divergence of the magnetization $\nabla \cdot m$ in [A/m²] by using the test data set from the Bavarian Molasse Basin

Taking a look at Fig. 6.25, we compare the decomposition of $B_{H^n_{\tau_j}}$ with the decomposition of the magnetic divergence $\nabla \cdot m$. The left side is the same as in Fig. 6.23. However, instead of the gradient, we apply the negative Laplacian in order to arrive at the decomposition of the magnetic divergence on the right side. It turns out that $\nabla \cdot m_{H^n_{\tau_j}}$, indeed, converges to $\nabla \cdot m$ as $j \to \infty$. Moreover, taking the difference $\nabla \cdot m_{H^n_{\tau_j}} - \nabla \cdot m_{H^n_{\tau_{j-1}}}$, the local transitional zones of the divergence $\nabla \cdot m$ become visible.

All in all, our mollifier approach represents a powerful tool, which is able to provide geological information in terms of the (divergence of the) magnetization of a particular region under consideration.

Chapter 7
Concluding Remarks

The aim of the book was to deal with the mathematical aspects of the ill-posed problem of transferring input magnetic information to output characteristics of the magnetization distribution field (or its divergence). Some essential features of the Biot–Savart law were recapitulated, and the magnetization distribution was explained in the framework of magnetostatic dipole-based potential theory. Different methodologies of multi-scale mollifier resolution of the inverse magnetometry problem were examined including their dependence on the data source. In this respect, three types of input information may be distinguished, namely internal (borehole), terrestrial (surface), and/or external (spaceborne) magnetic data sets. Singular integral theory oriented inversion of the volume dipole integral equation by Haar-type mollifiers was handled in a multi-scale framework to decorrelate specific signal signatures of the magnetization distribution with respect to inherently given geological constituents. Reproducing kernel Hilbert space regularization techniques were studied (together with their transition to certain mollifier variants) to provide deep magnetization distribution signatures by "downward continuation" from terrestrial and/or spaceborne data. It turned out that the mathematical applicability of mollifier techniques in exploration is of manifold advantage in methodological as well as geodisciplinary respects (see also the remarks in [45]).

7.1 Methodological Aspects

The methodological aspects of mollifier approximations in geoexploration can be characterized as follows:

- Mollifier methods can be used for a *multi-scale postprocessing* of already existing geological models. So, multi-scale postprocessing needs some knowledge of earlier geological work; its interpretation by mollifier techniques, however, offers new light in the interpretation of existing geological information by its

© The Author(s), under exclusive license to Springer Nature Switzerland AG 2021
C. Blick et al., *Inverse Magnetometry*, Lecture Notes in Geosystems Mathematics
and Computing, https://doi.org/10.1007/978-3-030-79508-5_7

decorrelation ability. In fact, a particular point for the realization of mollifier techniques as an important postprocessing procedure in exploration is the claim to enable a reduction of the project risk with low financial expenditure. In fact, to control the risk of success (cf. Fig. 7.26), one is well advised to consider cost-effective potential methods in exploration. Mollifier decorrelation provides helpful decision support from different synergetic aspects, e.g., to determine suitable natural fault zones and layers at sufficient depth in regions with anthropogenic characteristics.

- Another form of applicability is multi-scale mollifier *inversion* using spline and/or wavelet procedures as indicated in this chapter within the geomagnetic context. The essential calamity is the ill-posedness of the singular integral equation problem that needs to be inverted. In this respect, multi-scale "mollifier approximations" respresent scale-dependent "replacement solutions" of associated regular integration problems with the quality to come close to the solution of the real problem by a zooming-in procedure. Mollifier wavelet inversion constructions and resulting decorrelation methods as developed in this work provide novel methods, structures, and tools of far-reaching significance, and they represent a completely new approach to modern exploration. The essential

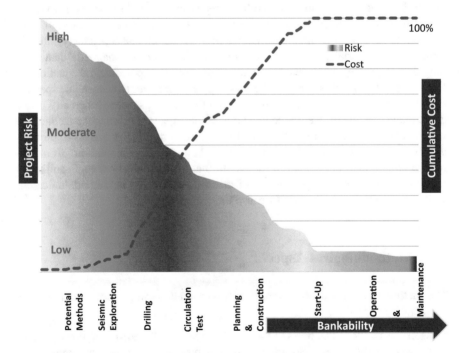

Fig. 7.26 Geothermal project risks, cumulative investment costs, and financial feasibility (illustration according to a template in the "Handbuch Tiefe Geothermie" [11–13] (see also the "Technical Report" of the Geothermal Energy Association (2011))

advantage of the mollifier tools is the inversion involving methods of sparse numerics.

7.2   Geodisciplinary Aspects

Gravimetric and magnetometric exploration aims at locating geothermal resources and water reservoirs, minerals, faults, and/or petroleum resources. Potential field surveys are relatively cheap, non-invasive, and nondestructive environmentally speaking. They are passive, i.e., no energy needs to be put into the ground in order to acquire data. The small portable instruments (gravimeter and magnetometer) also permit walking traverses. The goal of studying potential fields is to provide a better understanding of the subsurface geology. Gravity and magnetic prospecting is affected by the fact that the measured signal is a composite of the contributions from all depths. In fact, the uniqueness problem in gravimetry and magnetometry is the most serious obstacle. Superpositions over all depths can only be separated suitably if additional information is available, e.g., from earlier geological campaigns or borehole activities.

Tomographic and Scattering Mollifier Exploration   Not only gravimetry and magnetometry lead to potential methodologies, although the measuring instrumentarium of other methods is different from those we know for gravimetry and magnetometry. As a matter of fact, as we have seen, similar mollifier techniques as deduced for gravimetry and magnetometry can be developed for other problems in geomathematics, for example, acoustically and elastically based scattering (cf. [19, 20]). Moreover, poroelastic mollifier modeling (cf. [46] for some entry studies) seems to be a promising field in the near future.

In mathematical abstraction (see, e.g., [70]), the gravimetric and magnetometric mollifier approaches can be identified as prototypes for some other potential methods, which may be characterized in a condensed formulation as follows:

$$V(x) = \int_{\mathcal{B}} G(L; x, y) \, F(y) \, dy \tag{7.299}$$

involving the fundamental solution $G(L; x, y)$ corresponding to the differential operator L so that

$$F(x) = LV(x), \quad x \in \mathcal{B}, \tag{7.300}$$

where (in distributional sense)

$$L_x G(L; x, y) = \delta(x, y), \quad x \in \mathcal{B} \tag{7.301}$$

and

$$F(x) = \int_{\mathcal{B}} \delta(x, y) \, F(y) \, dy, \quad x \in \mathcal{B}. \tag{7.302}$$

In dependence on the specific choice of the operator L we are then led to the following exploration areas:

(i) *Gravitational modeling* (L (negative) Laplace operator)
(ii) *Geomagnetic modeling* (L (dipole-oriented) pre-Maxwell operators)
(iii) *Acoustic seismic tomography* (L (negative) Helmholtz operator)
(iv) *Elastic seismic tomography* (L Cauchy–Navier operator)
(v) *Acoustic scattering* (L time-dependent acoustic wave equation)
(vi) *Elastic scattering* (L time-dependent elastic wave equation)

7.3 Numerical Valuation

Multi-scale techniques using mollifier scaling and wavelet functions as constructed in this geomagnetically reflected contribution for explorational application, indeed, represent an outstanding methodology by its particular ability to elaborate specific features of the magnetization distribution by a "zooming-in" process. As an immediate consequence, multi-scale mollifier decorrelation may be rated without any doubt as a well-promising and far-reaching methodology in future magnetometric exploration.

However, there is also no doubt that each method in numerics has its own aim and character. In fact, it is the essence of any numerical realization that it becomes optimal only with respect to especially specified criteria. For example, radial basis functions do not show any directional modeling dependence. Trial functions with a local support corresponding to terrestrial-only data systems allow efficient and economical numerics for a near-Earth's surface domain, but if only terrestrial data are available, they do not enable a deeper modeling than the extension radius of the supports, also. In other words, for the numerical work to be done, we have to make an a priori choice that feature should be pointed out and preferably handled. In particular, for constructive approximation involving mollifier obligations, we have to reflect the different stages of space/frequency localization so that the modeling process can be performed adaptively under the localization requirements necessary and sufficient for geological interpretation.

Hence, from a numerical point of view, a positive decision in one direction often amounts to a negative implication in another direction. Ultimately, also because of the interdependence of the space/frequency localization caused by the validity of the uncertainty principle in peremptory terms (see, e.g., [45, 61]), it is impossible to provide a "cure-all methodology." This "sine qua non" ingredient in any constructive mathematical approximation method is also the reason, why we are indispensably led to essential difficulties in the mollifier approximation by scaling and wavelet potentials as proposed here:

- The choice of the type of the mollifier scaling function is an important problem in inverse magnetometry. In this chapter we restrict ourselves to locally supported trial functions such as Haar-type kernels, but also non-space-limited trial functions (e.g., Gauss-type functions) are applicable. In principle, all mollifier wavelet strategies are equivalent; however, the suitable computational compromise between mathematical rigor and practical relevance and applicability with respect to the specific geometry of the test area, data width, certain accuracy of the data, and occurring noise level and type is a task that should not be underestimated.

- In our approach described above, we are not confronted with a multi-scale procedure, for which subsequently each detail information guarantees an improvement to come closer and closer to the "real" solution. Well-posed problems usually do not require any stopping strategy of the multi-scale (zooming-in) process, since the data are (generally) given in a non-attenuated form. An algorithm establishing an approximate solution for the inverse magnetometry problem, however, has to take into account the requirement to stop at the right level of approximation in order to appropriately model the filtered occurrence of the measured potential data as smooth volume integral values. As a consequence, a missing stopping strategy in the multi-scale mollifier regularization usually produces considerable errors in the magnetization distribution even from extremely small errors in the potential measurements. In other words, there is a strong need for a relevant and mathematically motivated stopping strategy, which should not be underestimated as well.

In conclusion, a certain loss of information in numerical modeling of the magnetization distribution from (smooth) potential (integral) values is unavoidable. An algorithm establishing an approximate solution for the inverse magnetometric problem has to reflect the intention of exploration as good as possible. Moreover, geological *a priori* information today for the characterization of the right scaling and wavelet mollifier realizations for the "solution process" in inverse magnetometry as well as a mathematically validated stopping strategy in multi-scale mollifier regularization is an important challenge for future work.

All in all, it should be nonetheless pointed out that a mollifier methodology is able to determine both the signature information of the magnetization distribution contained in the smooth geomagnetic integral data and the band structures to detect the separating surfaces between differently magnetized geological strata via a decorrelation process.

List of Symbols

f, g .... Tensor-valued functions
$F|M, f|M, \mathbf{f}|M$ .... Restrictions to M
a .... Matrix in $\mathbb{R}^{q \times q}$
$\mathbf{a}^T$ .... Transposed of the matrix $\mathbf{t}$ in $\mathbb{R}^{q \times q}$
det **a** .... Determinant of the matrix $\mathbf{t}$ in $\mathbb{R}^{q \times q}$
i .... Identity tensor in $\mathbb{R}^{q \times q}$
0 .... Zero-tensor in $\mathbb{R}^{q \times q}$

dx .... Volume element (regarding the variable x)
$dS(\xi)$ .... Surface element (regarding the variable ξ)
$d\sigma(\xi)$ .... Curve element (regarding the variable ξ)

o, O .... (Conventional) Landau symbols

$\mathcal{B}$ .... Region in $\mathbb{R}^3$
$\overline{\mathcal{B}} = \mathcal{B} \cup \partial\mathcal{B}$ .... Topological closure of $\mathcal{B}$
$\partial\mathcal{B}$ .... Boundary of $\mathcal{B}$
$||\mathcal{B}||$ .... Volume of the region $\mathcal{B}$ in $\mathbb{R}^3$

$\alpha(x)$ .... Solid angle with respect to the surface $\partial\mathcal{B}$ at $x \in \mathbb{R}^3$

$\mathbb{S}^2$ .... Unit sphere around the origin 0 in $\mathbb{R}^3$
$\mathbb{S}^2_R$ .... Sphere with the radius R around 0 in the $\mathbb{R}^3$
$\mathbb{S}^2_R(y)$ .... Sphere with radius R around y in $\mathbb{R}^3$
$\mathbb{B}^3$ .... Unit ball around 0 in $\mathbb{R}^3$
$\mathbb{B}^3_R$ .... (Open) ball with radius R around 0 in $\mathbb{R}^3$
$\mathbb{B}^3_R(y)$ .... (Open) ball with radius R around y in $\mathbb{R}^3$
$\mathbb{B}^3_{R_0, R_1}(y) = \mathbb{B}^3_{R_1}(y) \backslash \overline{\mathbb{B}^3_{R_0}(y)}$ .... Spherical shell (ball ring) around y in $\mathbb{R}^3$
$\mathbb{B}^3_{R_0, R_1} = \mathbb{B}^3_{R_1} \backslash \overline{\mathbb{B}^3_{R_0}}$ .... Spherical shell (ball ring) around 0 in $\mathbb{R}^3$

$\frac{\partial}{\partial x_i}, i = 1, 2, 3,$ .... Partial derivatives in $\mathbb{R}^3$
∇ .... Gradient in $\mathbb{R}^3$
$\nabla \cdot$ .... Divergence in $\mathbb{R}^3$
L .... Curl gradient in $\mathbb{R}^3$
$L \cdot$ .... Curl divergence in $\mathbb{R}^3$
$\nabla \otimes \nabla$ .... Hesse-tensor in $\mathbb{R}^3$
Δ .... Laplace operator at $\mathbb{R}^3$

ξ, η, ζ .... Points of the unit sphere $\mathbb{S}^2$
∇^* .... Surface gradient on $\mathbb{S}^2$
$\nabla^* \cdot$ .... Surface divergence on $\mathbb{S}^2$
L^* .... Surface curl gradient on $\mathbb{S}^2$
$L^* \cdot$ .... Surface curl divergence on $\mathbb{S}^2$
Δ^* .... (Surface) Beltrami operator on $\mathbb{S}^2$

References

1. Alberto, P., Oliveira, O., Pais, M.A.: On the non-uniqueness of main geomagnetic field determined by surface intensity measurements: The Backus Problem. Geophys. J. Int. **159**, 548–554 (2004)
2. Alldredge, L.R., Hurwitz L.: Radial dipoles as the sources of the Earth's main magnetic field. J. Geophys. Res. **69**, 2631–2640 (1964)
3. Andres, G.: Fränkische Alb und Malmkarst des Molassebeckens. In: Grundwassergleichenkarte von Bayern 1:500,000 mit Erläuterungen. Schriftenreihe Bayer. Landesamt für Wasserwirtschaft, München. vol. 20, pp. 23–25 (1985)
4. Aronszajn, N.: Theory of reproducing kernels. Trans. Am. Math. Soc. **68**, 337–404 (1950)
5. Backus, G.E.: Non-uniqueness of the external geomagnetic field determined by surface intensity measurements. J. Geophys. Res. **75**, 6339–6341 (1970)
6. Backus, G.E.: Poloidal and toroidal fields in geomagnetic field modeling. Rev. Geophys. **24**, 75–109 (1986)
7. Backus, G.E., Gilbert, F.: Numerical applications of a formalism for geophysical inverse problems. Geophys. J. R. Astron. Soc. **13**, 247–276 (1967)
8. Backus, G.E., Gilbert, F.: The resolving power of gross Earth data. Geophys. J. R. Astron. Soc. **16**, 169–205 (1968)
9. Backus, G.E., Gilbert, F.: Uniqueness of the inversion of inaccurate gross Earth data. Philos. Trans. R. Soc. Lond. **226**, 123–197 (1970)
10. Backus, G.E., Parker, R., Constable, C.: Foundations of Geomagnetism. Cambridge University, Cambridge (1996)
11. Bauer, M., Freeden, W. Jacobi, H., Neu, T. (eds.): Handbuch Tiefe Geothermie. Springer, New York (2014)
12. Bauer, M., Freeden, W. Jacobi, H., Neu, T.: Energiewirtschaft 2014: Fakten und Chancen der Tiefen Geothermie, pp. 1–38. Springer Spektrum, Fachmedien Wiesbaden, Wiesbaden (2014)
13. Bauer, M., Freeden, W. Jacobi, H., Neu, T. (eds.): Handbuch Oberflächennahe Geothermie. Springer, New York (2018)
14. Bayer, M.: Geomagnetic Field Modeling from Satellite Data by First and Second Generation Wavelets, Ph.-D. thesis. University of Kaiserslautern, Geomathematics Group, Aachen (1999)
15. Bayer, M., Beth, S., Freeden, W.: Geophysical field modeling by multiresolution analysis. Acta Geod. Geoph. Hung. **33**, 289–319 (1998)
16. Bayer, M., Freeden, W., Maier, T.: A vector wavelet approach to iono- and magnetospheric geomagnetic satellite data. J. Atmos. Sol. Terr. Phys. **63**, 581–597 (2001)
17. Birner, J., Fritzer, T., Jodocy, M., Savvatis, A., Schneider, M., Stober, I.: Hydraulische Eigenschaften des Malmaquifers im Süddeutschen Molassebecken und ihre Bedeutung für die geothermische Erschließung. Z. Geol. Wiss. **40**, 133–156 (2012)

© The Author(s), under exclusive license to Springer Nature Switzerland AG 2021
C. Blick et al., *Inverse Magnetometry*, Lecture Notes in Geosystems Mathematics and Computing, https://doi.org/10.1007/978-3-030-79508-5

18. Blakely, R.J.: Potential Theory in Gravity and Magnetic Application. Cambridge University, Cambridge (1996)
19. Blick, C.: Multi-scale potential methods in geothermal research: decorrelation reflected post-processing and locally based inversion, Ph.-D. thesis. University of Kaiserslautern, Geomathematics Group, Kaiserslautern (2015)
20. Blick, C., Eberle, S.: Multi-scale density decorrelation by Cauchy–Navier wavelets. GEM Int. J. Geomath. **10**, 1–31 (2019)
21. Blick, C., Eberle, S.: A survey on multi-scale decorrelation for seismic data. GEM Int. J. Geomath. **12**, 16 (2021). https://doi.org/10.1007/s13137-021-00179-x.
22. Blick, C., Freeden, W., Nutz, H.: Innovative Explorationsmethoden am Beispiel der Gravimetrie und Reflexionsseismik. In: Bauer, M., Freeden, W., Jacobi, H., Neu, T. (eds.) Handbuch Oberflächennahe Geothermie, pp. 221–256. Springer Spektrum, Heidelberg (2018a)
23. Blick, C., Freeden, W., Nutz, H.: Gravimetry and exploration. In: Freeden, W., Nashed, M.Z. (eds.) Handbook of Mathematical Geodesy. Geosystems Mathematics, pp. 687–752. Springer, Birkhäuser (2018b)
24. Bosch, M., McGaughey, J.: Joint inversion of gravity and magnetic data under lithologic constraints. Leading Edge **20**, 877–881 (2001)
25. Bossavit, A.: Computational Eloctromagnetism. Academic Press, San Diego (1998)
26. Bullard E.C.: The magnetic field within the Earth. Proc. Camb. Phil. Soc. **A197**, 438–453 (1949)
27. Carmichael, R.S.: Magnetic properties of minerals and rocks. In: Carmichael, R.S. (ed.) Handbook of Physical Properties of Rocks, vol. II. CRC Press, Boca Raton (1982)
28. Chapman, S., Bartels, J.: Geomagnetism. Oxford University, Oxford (1940)
29. Claerbout J.F.: Fundamentals of geophysical data processing: with applications to Petroleum Prospecting. McGraw Hill, New York (1976)
30. Clark, D.A., Emerson, J.B.: Notes on rock magnetization characteristics in applied geophysical studies. Explor. Geophys. **22**, 448–455 (1991)
31. Davis, K., Li, Y., Fast solution of geophysical inversion using adaptive mesh, space-filling curves and wavelet compression. Geophys. J. Int. **185**, 157–166 (2011)
32. Dorsch, K., Pletl, C.: Bayerisches Molassebecken—Erfolgsregion der Tiefengeothermie in Mitteleuropa. Geothermische Energie **73**, 14–18 (2012)
33. Eggermont, P.N., LaRiccia, V., Nashed, M.Z.: Noise models for ill-posed problems. In: W. Freeden, Z. Nashed, T. Sonar (eds.) Handbook of Geomathematics, 2nd edn., vol. 2, pp. 1633–1658. Springer, New York (2015)
34. Elsasser, W.M.: Induction effects in terrestrial magnetism. I. Theory. Phys. Rev. **69**, 106–116 (1946)
35. Engl, H.: Integralgleichungen. Springer Lehrbuch Mathematik, Wien (1997)
36. Engl, H.W., Hanke, M., Neubauer, A.: Regularization of Inverse Problems. Kluwer Academic Publisher, Dordrecht (1996)
37. Engl, H., Louis, A.K., Rundell, W. (eds.) Inverse problems in geophysical applications. SIAM, Philadelphia (1997)
38. Eskola, L.: Geophysical Interpretation Using Integral Equations. Chapman and Hall, London (1992)
39. Farquharson, C.G., Oldenburg, D.W., Non-linear inversion using general measures of data misfit and model structure. Geophys. J. Int. **134**, 213–227 (1998)
40. Fehlinger, T.: Multiscale formulations for the disturbing potential and the deflections of the vertical in locally reflected Physical Geodesy, Ph.-D. thesis. University of Kaiserslautern, Geomathematics Group, Kaiserslautern (2009)

41. Freeden, W.: On approximation by harmonic splines. Manuscr. Geodaet. **6**, 193–244 (1981)
42. Freeden, W.: Multiscale Modeling of Spaceborne Geodata. Teubner, Stuttgart (1999)
43. Freeden, W.: Geomathematik, was ist das überhaupt? Jahresber. Deutsch. Math. Vereinigung (DMV) **111**, 125–152 (2009)
44. Freeden, W.: Geomathematics: Its Role, its Aim, and its Potential. In: Freeden, W., Nashed, Z., Sonar, T. (eds.) Handbook of Geomathematics, 2nd edn., vol. 1, pp. 3–78 Springer, New York (2015)
45. Freeden, W.: Decorrelative Mollifier Gravimetry–Basics, Concepts, Examples and Perspectives. Geosystems Mathematics, Birkhäuser (2021)
46. Freeden, W., Bauer, M.: Dekorrelative Gravimetrie—Ein innovativer Zugang in Exploration und Geowissenschaften. Springer Spektrum, Berlin (2020)
47. Freeden, W., Blick, C.: Signal decorrelation by means of multiscale methods. World Min. **65**, 304–317 (2013)
48. Freeden, W., Gerhards, C.: Poloidal and toroidal modeling in terms of locally supported vector wavelets. Math. Geosci. **42**, 817–838 (2010)
49. Freeden, W., Gerhards, C.: Geomathematically Oriented Potential Theory. CRC Press/Taylor and Francis, Boca Raton (2013)
50. Freeden, W., Gutting, M.: Special Functions of Mathematical (Geo)Physics. Birkhäuser, Basel (2013)
51. Freeden, W., Gutting, M.: Integration and Cubature Methods. Chapman and Hall/CRC Press, Boca Raton/New York (2018)
52. Freeden, W., Maier, T.: On multiscale denoising of spherical functions: basic theory and numerical aspects. Electron. Trans. Numer. Anal. (ETNA) **14**, 40–62 (2002)
53. Freeden, W., Maier, T.: Spectral and multiscale signal-to-noise thresholding of spherical vector fields. Comput. Geosci. **7**, 215–250 (2003)
54. Freeden, W., Mayer, C.: Wavelets generated by layer potentials. Appl. Comput. Harm. Anal. (ACHA) **14**, 195–237 (2003)
55. Freeden, W., Michel, V.: Multiscale Potential Theory (with Applications to Geoscience). Birkhäuser, Boston (2004)
56. Freeden, W., Nashed, M.Z.: Ill-posed problems: operator methodologies of resolution and regularization approaches. In: Freeden, W. and Nashed, M.Z. (eds.) Handbook of Mathematical Geodesy. Geosystems Mathematics, pp. 201–314. Springer/Birkhäuser, Basel/New York (2018a)
57. Freeden, W., Nashed, M.Z.: Inverse gravimetry as an ill-posed problem in Mathematical Geodesy. In: Freeden, W., Nashed, M.Z. (eds.) Handbook of Mathematical Geodesy. Geosystems Mathematics, pp. 641–685. Springer, Basel (2018b)
58. Freeden, W., Nashed, M.Z.: Inverse gravimetry: background material and multiscale mollifier approaches. GEM Int. J. Geomath. **9**, 199–264 (2018c)
59. Freeden, W., Nashed, M.Z.: Operator-theoretic and regularization approaches to ill-posed problems. GEM Int. J. Geomath. **9**, 1–115 (2018d)
60. Freeden, W., Nashed, M.Z.: Inverse gravimetry: density signatures from gravitational potential data. In: Freeden, W., Rummel, R. (eds.) Handbuch der Geodäsie, Mathematische Geodäsie/Mathematical Geodesy, pp. 969–1052. Springer Spektrum, Heidelberg (2020)
61. Freeden, W., Nashed, M.Z., Schreiner, M.: Spherical Sampling. Geosystems Mathematics, Springer, Basel (2018)
62. Freeden, W., Nutz, H.: Mathematik als Schlüsseltechnologie zum Verständnis des Systems "Tiefe Geothermie". Jahresber. Deutsch. Math. Vereinigung (DMV) **117**, 45–84 (2015)
63. Freeden, W., Nutz, H.: Inverse Probleme der Geodäsie: Ein Abriss mathematischer Lösungsstrategien. In: Freeden, W., Rummel, R. (eds.) Handbuch der Geodäsie, Mathematische Geodäsie/Mathematical Geodesy, pp. 65–90. Springer Spektrum, Heidelberg (2020)
64. Freeden, W., Schreiner, M.: Local multiscale modeling of geoid undulations from deflections of the vertical. J. Geodesy **79**, 641–651 (2006)
65. Freeden, W., Schreiner, M.: Spherical Functions of Mathematical Geosciences: A Scalar, Vectorial, and Tensorial setup. 1st edn., Springer, Heidelberg (2009)

66. Freeden, W., Schreiner, M.: Mathematical Geodesy: Its role, its aim, and its potential. In: Freeden, W., Rummel, R. (eds.) Handbuch der Geodäsie, Mathematische Geodäsie/Mathematical Geodesy, pp. 3–64. Springer Spektrum, Heidelberg (2020)

67. Freeden, W., Witte, B.: A combined (spline-)interpolation and smoothing method for the determination of the gravitational potential from heterogeneous data. Bull. Géod. **56**, 53–62 (1982)

68. Freeden, W., Nashed, Z., Sonar, T. (eds.): Handbook of Geomathematics, 2nd edn., vol. 1,2, and 3. Springer, New York (2015)

69. Freeden, W., Nashed, M.Z. (eds.): Handbook of Mathematical Geodesy. Geosystems Mathematics, Springer/Birkhäuser, New York (2018)

70. Freeden, W., Heine, C., Nashed M.Z.: An Invitation to Geomathematics. Lecture Notes in Geosystem Mathematics and Computing (2019)

71. Freeden W., Nutz H., Rummel R., Schreiner M.: Satellite gravity gradiometry (SGG): Methodological foundation and geomathematical advances. In: Freeden, W., Rummel, R. (eds.) Handbuch der Geodäsie, Mathematische Geodäsie/Mathematical Geodesy, pp. 1185–1256. Springer Spektrum, Heidelberg (2020)

72. Friedrichs, K.O.: Die Identität schwacher und starker Erweiterungen von Differentialoperatoren. Trans. Am. Math. Soc. **55**, 132–151 (1944)

73. Gauss, C.F.: Allgemeine Theorie des Erdmagnetismus. Resultate aus den Beobachtungen des magnetischen Vereins, Göttingen (1838)

74. Gerhards, C.: Spherical Multiscale Methods in Terms of Locally Supported Wavelets: Theory and Application to Geomagnetic Modeling, Ph.-D. thesis. University of Kaiserslautern, Geomathematics Group, Kaiserslautern (2011)

75. Glaßmeier, K.H., Soffel, H., Negendank, J. (eds.) Geomagnetic field variations. In: Advances in Geophysical and Environmental Mechanics and Mathematics. Springer, Berlin (2009)

76. Glaßmeier, K.H., Soffel, H., Negendank, J.: The geomagnetic field. In: Glaßmeier, K.H., Soffel, H., Negendank, J. (eds.) Geomagnetic field variations. Advances in Geophysical and Environmental Mechanics and Mathematics, pp. 1–23. Springer, Berlin (2009)

77. Gubbins, D., Herrero-Bervera, E. (eds.): Encyclopedia of Geomagnetism and Paleomagnetism. Springer, Dordrecht (2007)

78. Gui, Y.F., Dou, W.B.: A rigorous and completed statement on Helmholtz theorem. Prog. Electromagn. Res. (PIER) **69**, 287–304 (2007)

79. Haar, A.: Zur Theorie der orthogonalen Funktionensysteme. Math. Ann. **69**, 331–371 (1910)

80. Hadamard, J.: Sur les problèmes aux dérivés partielles et leur signification physique. Princeton Univ. Bull. **13**, 49–52 (1902)

81. Hadamard, J.: Lectures on the Cauchy-problem in linear partial differential equations. Yale University, New Haven (1923)

82. Hauser, W.: Introduction to the Principles of Electromagnetism. Addison-Wesley, Boston (1971)

83. Hulot, G., Finlay, C.C., Constable, C., Olsen, N., Mandea, M.: The magnetic field of planet earth. Space Sci. Rev. **152**, 159–222 (2010)

84. Hulot, G., Olsen, N., Sabaka, T.J., Fournier, A.: The Present and Future Geomagnetic Field (2015). DOI: 10.1016/B978-0-444-53802-4.00096-8

85. Jackson, J.D.: Classical Electrodynamics. Wiley, New York (1998)

86. Jacobs, F., Meyer, H.: Geophysik–Signale aus der Erde. Teubner, Leipzig (1992)

87. Kellogg, O.D.: Foundations of Potential Theory. Frederick Ungar Publishing Company, New York (1929)

88. Kono, M. (ed.): Geomagnetism, Treatise on Geophysics, vol. 5. Elsevier, Amsterdam (2009)

89. Langel, R.A., Hinze, W.J.: The magnetic field of the Earth's lithosphere: the satellite perspective. Cambridge University, Cambridge (1998)

90. Lelièvre, P.G.: Integrating geologic and geophysical data through advanced constrained inversions, Ph.-D. thesis. The University of British Columbia, Vancouver (2009)

91. Lelièvre, P.G., Farquharson, C.G.: Gradient and smoothness regularization operators for geophysical inversion on unstructured meshes. Geophys. J. Int. **195**, 330–341 (2013)

92. Lelièvre, P.G., Oldenburg, D.W.: Magnetic forward modeling and inversion for high susceptibility. Geophys. J. Int. **166**, 79–90 (2006)
93. Leweke, S., Michel, V., Telschow, R.: On the uniqueness of gravitational and magnetic field data inversion. In: Freeden, W., Nashed, M.Z. (eds.): Handbook of Mathematical Geodesy. Geosystems Mathematics, pp. 883–920. Springer International Publishing, Birkhäuser, New York (2018)
94. Lima, E.A., Irimia, A., Wikswo, J.P.: The magnetic inverse problem. In: Clarke, J., and Braginski, A.E. (eds.) The SQUID Handbook, vol. II. WILEY-VCH, Weinheim (2006)
95. Louis, A.K., Maass, P.: A mollifier method for linear equations of the first kind. Inverse Prob. **6**, 427–440 (1990)
96. Lowes, F.J.: Spatial power spectrum of the main geomagnetic field, and extrapolation to the core. Geophys. J. R. Astron. Soc. **36**, 717–730 (1974)
97. Lühr, H., Korte, M., Mandea, M.: The recent geomagnetic field and its variations. In: Glaßmeier, K.H., Soffel, H., Negendank, J. (eds.) Geomagnetic Field Variations. Advances in Geophysical and Environmental Mechanics and Mathematics, pp. 25–64. Springer, Berlin (2009)
98. Lukyanenko, D., Yagola, A.: Some methods for solving of 3D inverse problem of magnetometry. Eurasian J. Math. Comput. Appl. **4**, 4–14 (2016)
99. Maier, T.: Multiscale geomagnetic field modeling from satellite data: theoretical aspects and numerical applications, Ph.-D. thesis. University of Kaiserslautern, Geomathematics Group, Kaiserslautern (2002)
100. Maier, T.: Wavelet-Mie-representation for solenoidal vector fields with applications to ionospheric geomagnetic data. SIAM J. Appl. Math. **65**, 1888–1912 (2005)
101. Mareš, S., Tvrdý, M.: Introduction to Applied Geophysics. Kluwer, Dordrecht (1984)
102. Martensen, E.: Potentialtheorie. In: Leitfäden der Angewandten Mathematik und Mechanik, Bd., vol. 12. Teubner, Leipzig (1968)
103. Martin, G.S., Marfurt, K.J., Larsen, S.: Marmousi-2: An updated model for the investigation of AVO in structurally complex areas. In: Proceedings, Society of Exploration Geophysicists Annual Meeting, Salt Lake City (2002)
104. Martin, M.S., Wiley, R., Marfurt, K.J.: Marmousi-2: An elastic upgrade for Marmousi. Lead. Edge **25**, 156–166 (2006)
105. Mauersberger, P.: Das Mittel der Energiedichte des geomagnetischen Hauptfeldes an der Erdoberfläche und seine säkulare Änderung. Gerlands Beiträge zur Geophysik **65**, 207–215 (1956)
106. Mayer, C.: Wavelet modeling of ionospheric currents and induced magnetic fields from satellite data, Ph.-D. thesis. University of Kaiserslautern, Geomathematics Group, Kaiserslautern (2003)
107. Mayer, C.: Wavelet modelling of the spherical inverse source problem with application to geomagnetism. Inverse Prob. **20**, 1713–1728 (2004)
108. Mayer, C., Maier, T.: Separating inner and outer Earth's magnetic field from CHAMP satellite measurements by means of vector scaling functions and wavelets. Geophys. J. Int. **167**, 1188–1203 (2006)
109. Menke, W.: Geophysical Data Analysis: Discrete Inverse Theory. Academic Press, Orlando (1984)
110. Meyer, R.K.F., Schmidt-Kaler, H.: Jura. Bayerisches Geologisches Landesamt (eds.), Erläuterungen zur Geologischen Karte von Bayern 1:500 000 (4. Aufl.) München, pp. 90–111 (1996)
111. Michel, V.: A multiscale approximation for operator equations in separable Hilbert spaces—case study: reconstruction and description of the Earth's interior. University of Kaiserslautern, Geomathematics Group, Habilitation Thesis (2002)
112. Moeck, I.: Vorreiter der weltweiten Entwicklung geothermischer Ressourcen: Das Bayerische Molassebecken. GTE **81**, 22–26 (2015)
113. Möhringer, S.: Decorrelation of gravimetric data, Ph.-D. thesis. University of Kaiserslautern, Geomathematics Group, Kaiserslautern (2014)

114. Morse, P.M., Feshbach, H.: Methods of Theoretical Physics. McGraw-Hill, New York (1953)
115. Müller, C.: Foundations of the Mathematical Theory of Electromagnetic Waves. Springer, Berlin (1969)
116. Morse, P.M., Feshbach, H., Hill, E.L.: Methods of Theoretical Physics. McGraw-Hill, New York (1953)
117. Nashed, M.Z.: Generalized inverses, normal solvability and iteration for singular operator equations. In: Rall, L.B. (ed.) Nonlinear Functional Analysis and Applications, pp. 311–359. Academic Press, New York (1971)
118. Nashed, M.Z.: Aspects of generalized inverses in analysis and regularization. In: Generalized Inverses and Applications, pp. 193–244. Academic Press, New York (1976)
119. Nashed, M.Z.: Operator-theoretic and computational approaches to ill-posed problems with applications to antenna theory. IEEE Trans. Antennas Propag. **29**, 220–231 (1981)
120. Nashed, M.Z.: A new approach to classification and regularization of ill-posed operator equations, inverse and ill-posed problems. In: Engl, H.W. and Groetsch, C.W. (eds.) Notes and Reports in Mathematics in Science and Engineering, vol. 4. Academic Press, New York (1987)
121. Nashed, M.Z.: In: Siddiqi, A.H., Singh, R.C. Manchanda, P. (eds.) Inverse problems, moment problems, and signal processing: Un Menage a Trois. Mathematics in science and technology, pp. 1–19. World Scientific, New Jersey (2010)
122. Nashed, M.Z., Scherzer, O.: Inverse Problems, Image Analysis and Medical Imaging (Contemporary Mathematics), vol. 313. American Mathematical Society, Providence (2002)
123. Nashed, M.Z., Votruba, F.G.: A unified operator theory of generalized inverses. In: Nashed, M.Z. (ed.) Generalized Inverses and Applications, pp. 1–109. Academic Press, New York (1976)
124. Nashed, M.Z., Walter, G.G.: General sampling theorems for functions in reproducing kernel Hilbert space. Math. Control Signals Syst. **4**, 363–390 (1991)
125. Nashed, M.Z., Walter, G.G.: Reproducing kernel Hilbert space from sampling expansions. Contemp. Math. **190**, 221–226 (1995)
126. Olsen, N., Glassmeier, K.-H., Jia, X.: Separation of the magnetic field into external and internal parts. Space Sci. Rev. **152**, 159–222 (2010)
127. Olsen, N., Hulot, G., Sabaka, T.J.: Sources of the geomagnetic field and the modern data that enable their investigation. In: Freeden, W., Nashed, Z., Sonar, T. (eds.) Handbook of Geomathematics, 1st edn, vol. 1, pp. 106–124. Springer, London (2010)
128. Parker, R.L.: The inverse problem of electromagnetic induction: Existence and construction of solutions based on incomplete data. J. Geophys. Res. **85**, 3321–3328 (1980)
129. Parkinson, W.D.: Introduction to Geomagnetism. Scottish Academic Press, Edinburgh (1983)
130. Paterson, N.R., Reeves, C.V.: Applications of Gravity and Magnetic Surveys: The State-of-the-Art. Geophysics **50**, 2558–2594 (1985)
131. Rivas, J.: Gravity and magnetic methods. In: Short Course on Surface Exploration for Geothermal Resources, Organized by UNU-GTP and LaGeo, in Ahuachapan and Santa Tecla, El Salvador, pp. 1–13 (2009)
132. Saltus, R.W., Blakely, R.J.: Unique geologic insights from "non-unique" gravity and magnetic interpretation. GSA Today **21**, 4–11 (2011)
133. Shure, L., Parker, R.L., Backus, G.E.: Harmonic splines for geomagnetic modelling. Phys. Earth Planet. Inter. **28**, 215–229 (1982)
134. Shure, I., Parker, R.L., Langel, R.A.: A preliminary harmonic spline model from Magsat data. J. Geophys. Res. **90**, 11505–11512 (1985)
135. Sprössig, W.: On Helmholtz decompositions and their generalizations—An overview. Math. Meth. Appl. Sci. **33**, 374–383 (2010)
136. Stober, I., Jodocy, M.: Geothermische Nutzhorizonte im westlichen Teil des Süddeutschen Molassebeckens. Z. Geol. Wiss. **39**, 161–172 (2011)
137. Strakhov, V.N.: Solution of linear inverse problems of gravimetry and magnetometry. Dokl. Akad. Nauk SSSR **310**, 1358–1362 (1990)
138. Turcotte, D.L., Schubert, G.: Geodynamics. Cambridge University, Cambridge (2001)

139. Valenta, J.: Introduction to Geophysics–Lecture Notes. Czech Republic Development Cooperation, Prague (2015)
140. Versteeg, R.: The Marmousi experience: velocity model determination on a synthetic complex data set. Lead. Edge **13**, 927–936 (1994)
141. Vogt, J., Sinnhuber, M., Kallenrode, M.B.: Effects of geomagnetic variations on system Earth. In: Glaßmeier, K.H., Soffel, H., Negendank, J. (eds.) Geomagnetic Field Variations. Advances in Geophysical and Environmental Mechanics and Mathematics, pp. 159–208. Springer, Berlin (2009)
142. Walter, W.: Einführung in die Potentialtheorie. In: BI Hochschulskripten, pp. 765/765a (1971)
143. Walter, R.: Geologie von Mitteleuropa. 5. Aufl., E. Schweizerbart'sche Verlagsbuchhandlung, Germany (1995)
144. Wapler, C., Leupold, J., Dragonu, I., von Elverfeld, D., Zaitsev, M., Wallrabea, U.: Magnetic properties of materials for MR engineering, micro-MR and beyond. J. Magn. Reson. **242**, 233–242 (2014)
145. Wikswo J.P.: The magnetic inverse problem for NDE. In: Weinstock, H. (ed.) SQUID Sensors: Fundamentals, Fabrication and Applications. NATO ASI Series (Series E: Applied Sciences), vol. 329. Springer, Dordrecht (1996)
146. Wikswo J.P., Ma, Y.P., Sepulveda, N.G., Tan S. Thomas I.M., Lauder, A.: Magnetic susceptibility imaging for nondestructive evaluation (using SQUID magnetometer). IEEE Trans. Appl. Supercond. **3**(1), 1995–2002 (2002)
147. Wolf, K.: Multiscale modeling of classical boundary value problems in physical geodesy by locally supported wavelets, Ph.-D. thesis. University of Kaiserslautern, Geomathematics Group, Kaiserslautern (2009)
148. Zidarov, D.P.: Conditions for Uniqueness of Self-Limiting Solutions of the Inverse Problems. Comptes rendus de l'Académie bulgare des Sci. **39**, 57–60 (1986)

Index

© The Author(s), under exclusive license to Springer Nature Switzerland AG 2021
C. Blick et al., *Inverse Magnetometry*, Lecture Notes in Geosystems Mathematics
and Computing, https://doi.org/10.1007/978-3-030-79508-5

Printed in the United States
by Baker & Taylor Publisher Services